Practical
Design Control
Implementation
for
Medical Devices

...

Jose M. Justiniano
Venky Gopalaswamy

CRC Press
Taylor & Francis Group
Boca Raton London New York

CRC Press is an imprint of the
Taylor & Francis Group, an **informa** business

CRC Press
Taylor & Francis Group
6000 Broken Sound Parkway NW, Suite 300
Boca Raton, FL 33487-2742

First issued in paperback 2019

© 2003 by Taylor and Francis Group, LLC
CRC Press is an imprint of Taylor & Francis Group, an Informa business

No claim to original U.S. Government works

ISBN-13: 978-1-57491-127-5 (hbk)
ISBN-13: 978-0-367-39538-4 (pbk)
Library of Congress Card Number 2002013391

Library of Congress Cataloging-in-Publication Data

Justiniano, Jose M.
    Practical design control implementation for medical devices / Jose M. Justiniano and Venky Gopalaswamy.
       p. cm.
     Includes bibliographical references and index.
     ISBN 1-57491-127-9 (alk. paper)
     1. Medical Instruments and apparatus—Design and construction. 2. Medical instruments and apparatus—Quality control. I. Gopalaswamy, Venky. II. Title.
     [DNLM: 1. Equipment Design. 2. Equipment and Supplies—standards. 3. Quality Control.
W 26 J96p 2002]

R856.J87 2002
610´ .28´4—dc21   2002013391

Visit the Taylor & Francis Web site at
http://www.taylorandfrancis.com

and the CRC Press Web site at
http://www.crcpress.com

CONTENTS

PART ONE

Practical Design Control Implementation

INTRODUCTION

The complexity of applications and technologies in the medical device industry is enormous. Medical devices range from simple handheld tools to complex computer-controlled surgical machines, from implantable screws to artificial organs, from blood-glucose test strips to diagnostic imaging systems and laboratory test equipment. Given this complexity, it is no surprise that these devices are designed and manufactured by companies that not only vary in size and structure, but use different methods of design and development as well as management.

Technological advances allow the planning, design, manufacture, operation, and maintenance of a variety of medical device components and systems to be performed every day with great efficiency. When these products and systems fail, however, the result could range from inconvenience and irritation to a critical impact on society. One of the major concerns of users of products and systems is their reliability and availability. Needless to say, because high reliability and availability will result in better customer satisfaction, the current trend in product and system design is toward attaining these goals. Advances in technology have resulted in better manufacturing processes, production control, product design, and so on, and have enabled engineers to design, manufacture, and build reliable components and systems.

Until few years ago, a newly hired professional in the medical device or pharmaceutical industry was typically told that Good Manufacturing Practice (GMP) merely involved meeting specifications and following written procedures. The job of the quality professional was to be a "watchdog" of all documentation that support the manufacturing of the product (it still is, especially in the pharmaceutical arena). If manufacturing operations had to change a specification, no information was available to connect the specification to customer requirements or to address the consequences of such proposed changes in the field. It was simply not acceptable practice to review or challenge the design engineering group's decision to set the specification in a certain way.

Through our combined experience in the automotive, telecommunications, electronics, and medical device industries, we know that such connections and their formal documentation is just "commonsense engineering design work." The medical device industry—especially those medical device companies that are ISO 9001 certified—is fast learning what the automotive and aerospace industries already knew:

- The key to effective design and development of a product is to utilize an effective design control process.

- Using design control process and supporting systems will consistently result in products that are most likely to meet customer and regulatory requirements.

For all medical device manufacturers interested in marketing their products in the United States, the time to adopt better product design and development practices is now, especially because the grace period for implementing the Food and Drug Admininstration (FDA) design control guidelines ended on June 1, 1998. This means that a medical device manufacturer can now be cited via 483s, warning letters, or other FDA enforcement actions for failing to comply with the design control requirements of the Quality System Regulation. The encouraging news is that of the 582 design control inspections conducted by the FDA between June 1, 1997, and June 1, 1998, 355 (60 percent) of the devices covered were subject to design controls, and 530 (91 percent) of the firms had established design control procedures.

Despite this encouraging news, some firms might be in compliance by investing heavily in consultants as well as internal resources without obtaining the expected level of understanding required to implement design control guidance. Our strong belief is that canned software programs and even "quality systems" can be bought without achieving the essence or intent of the design control regulation. Compliance to design control guidance through established procedures is basic and will certainly have positive effects on traditionally uncontrolled disciplines in the design and development of medical devices.

As part of its focus on design control enforcement, the October 1999 issue of "The Silver Sheet" presented change control problems cited in 483s issued between June 1, 1998, and September 30, 1999. The top 12 deficiencies are cited in Table I.1 and shown graphically in Figure I.1.

The management and technical personnel of those firms that spent a great deal of time and resources might still be asking themselves the same questions:

- What is required in design control?

- What sections of the guidance document are applicable to us?

- Where is the line between process and product development?

TABLE I.1   Top 12 Design Deficiencies Cited by the FDA

Design Deficiency [CFR No.]	Percentage of 483 Design Control Observations	Number of Design Control Observations
Inadequate design and development plan (820.30[b])	10.8	56
Inadequate design history file (830.20[j])	8.1	42
Lack of design control system (820.30[a])	5.8	30
Inadequate procedure for design changes (820.30[i])	5.6	29
No procedure for design changes (820.20[i])	5.4	28
Failure to verify or validate design changes (820.20[i])	4.2	22
Failure to document design changes or incomplete documentation (820.30[i])	3.7	19
No procedure for design transfer (820.34[h])	2.9	15
Inadequate procedure for design input (820.30[c])	2.7	14
Inadequate procedure to validate device design (820.30[g])	2.7	14
No procedure to validate device design (820.30[g])	2.7	14
Failure to include a mechanism for addressing incomplete, ambiguous, or conflicting requirements (820.30[c])	2.5	13

Adapted from "The Silver Sheet." Copyright 1999 F-D-C Reports, Inc. Reprinted with permission.

- Will the FDA design control requirements apply to a manufacturing facility?

- What is "verification," and what is its connection to "validation"?

- What shall we do with existing products?

- During a pilot or qualification run, are we validating the design of the product or the process?

The positive effects of successful design control implementation can be very evident: fewer customer complaints and medical device reports (MDRs), more satisfied customers, and faster time to market. In addition, manufacturing operations should see fewer "deviations," fewer "nonconforming" reports, less scrap, and fewer retests or other "re-do" operations before shipping the products. Through the implementation of design control guidelines, the FDA has given the industry a compelling reason to increase its markets or market share by improving quality. Those companies that do not approach design control with enthusiasm or strength of conviction may eventually realize their weak positions among competitors.

Medical device companies that are in the process of true implementation of design control must realize that their industry is, as mentioned earlier, very complex and "high tech." The challenges posed by the elements involving life sciences, engineering,

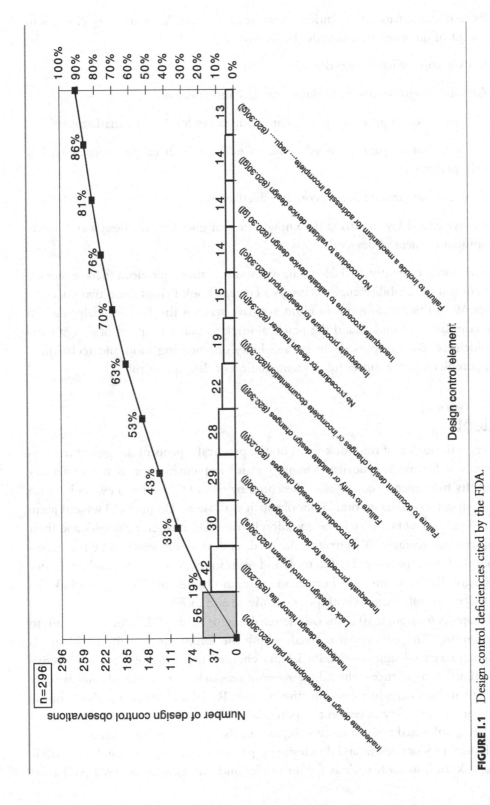

FIGURE I.1 Design control deficiencies cited by the FDA.

Source: Adapted from "The Silver Sheet." Copyright 1999 F-D-C Reports, Inc. Reprinted with permission.

statistics, and mathematics, to mention a few, are enormous. A good reality check is to answer a set of questions that include the following:

■ Do we know what we are doing?

■ Are our design control procedures practical and effective?

■ Can we predict our device performance and safety level in the marketplace?

■ Are our control parameters adequate to evaluate internal processes as well as field performance?

■ How can we measure design control effectiveness?

■ Are we effectively capturing the knowledge that goes into the design and development of medical devices?

The books currently available to answer some of these questions focus either on design control or reliability engineering. No current book brings these two concepts together. With these realizations in mind, we have written this book to help medical device companies to understand the practical implementation aspects of design control guidelines. We also present device reliability engineering tools and techniques, which complement the successful implementation of design control.

Content

Chapters 1 through 5 of this book focus on the practical aspects of design control implementation for medical devices. Design control implementation is not extremely difficult. Its fundamental concepts as presented by the FDA are not new and are no different than those already found in books such as those by Deming (1986) and Juran (1992). What is new, however, is the adoption by the FDA of such principles and their consequent enforcement. Also new to the medical device industry will be the different technical concepts as well as the tools and techniques necessary to implement design control. Every attempt has been made in each chapter of this book to link the FDA design control elements with practical advice and insight.

Chapters 6 through 10 focus on the reliability of medical devices and its link to design control. This concept is utilized during the entire product life cycle and is complementary to design control. Reliability engineering characterizes the objective of any engineering science—the achievement of products that are reliable and are easy to operate and maintain in a cost-effective manner. Reliability analysis is also a fundamental part in any safety assessment of potentially dangerous products, which are currently being subjected to even more stringent regulations and public attention.

Typical product design and development phases with the corresponding reliability and risk analysis tools such as failure modes and effects analysis (FMEA), hazard analysis, reliability planning, reliability prediction, design verification, and validation are presented with easy-to-understand examples. The linkage of reliability engineering to design control is presented throughout to help the reader understand how these concepts are interrelated. We sincerely think that this book will help medical device companies, both small and large, to successfully implement design control and reliability engineering to not only meet the FDA guidelines but also meet or exceed customer expectation.

Intended Audience

We wrote this book with all levels of management and technical personnel in mind. Whereas the Introduction and Chapter 1 can give top management a macro view of what it takes to develop and design medical devices per the Quality System Regulation, Chapters 2 through 5 are more tactical in nature. These chapters, however, provide useful information to top management regarding what kinds of human resources and technical knowledge are needed.

Chapters 6 and beyond are specifically aimed at technical personnel involved with product design and design changes. A reliability engineering background is not necessary because the material presented is very simple and can be seen as an introduction.

Another way to look at the chapters is as follows: Chapters 1 through 5 address the issue of how to reliably design and develop medical devices, whereas Chapters 6 through 11 address the issue of how to design and develop reliable medical devices.

This book is useful whether the reader is interested in strategic or tactical implementation of design control for medical device design. The main purpose of this book is to expose the medical device industry audience to a topic that is very applicable to design control and validation. Top management is encouraged to read Chapters 6 through 11 to assess the resources needed to design reliable medical devices.

CHAPTER
ONE

Motivation for Design Control and Validation

The motivation for controlling the process of designing medical devices is twofold. First, the obvious motivation is that the FDA included such control as part of its Quality System Regulation. The second and better motivation is that such a set of disciplines can improve business for any manufacturer. This chapter aims to impart an understanding of what the FDA and its Center for Devices and Radiological Health (CDRH) are seeking with design controls and validations. At the same time, the chapter addresses the business benefits beyond being in compliance with the regulation.

Motivation from the FDA and the CDRH

CDRH data analysis of the proportion of quality problems resulting in recalls between 1985 and 1989 revealed that approximately 50 percent were due to Good Manufacturing Practice (GMP) and 45–50 percent were preproduction (with no legal authority over the design process) (FDA CDRH "Human Factors Implications"). Half of the GMP problems were related to the manufacturer's ability to control the manufacturing process.[1] Both percentages suggest the need for a greater emphasis on design control and process validation.

A January 1990 FDA study, "Device Recalls: A Study of Quality Problems," based on data between October 1983 and September 1989, found that 44 percent of the quality problems leading to voluntary recalls were attributable to device design deficiencies (FDA CDRH 1990). The FDA deemed that percentage to be unacceptable and, after reviewing the data, determined that most of these design problems were preventable. Another FDA study showed that design controls can reduce design problems by 73 percent, which is equivalent to preventing 44 deaths per year (FDA

---

1. Defining what has to be controlled is a responsibility of the design and development team, which is supposed to "develop the process." Later in this book, we elaborate on this topic.

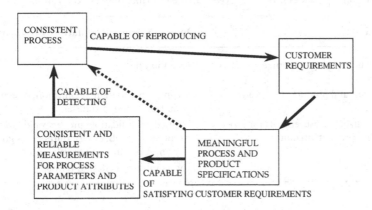

FIGURE 1.1   Practical business needs.

CDRH 1996a). The FDA understands that ineffective and/or unsafe devices can be the result of informal device development. In practical terms, informality is the lack of a disciplined approach.

The Safe Medical Devices Act of 1990 added design control (820.30) requirements to the GMP in section 520(f) of the act. Prior to 1990, the FDA used the term "preproduction" in an attempt to go beyond the typical "cGMPs" (current GMPs). Additionally, since 1990, the medical device industry has seen heavier scrutiny on compliance with process validation (820.75) and the effectiveness of corrective and preventive action (CAPA; 820.70[a] and [b], and 820.100[2] and [3]).

The Business Motivation

Any business that develops and manufactures any product has a set of practical needs. Such a set of needs becomes more relevant as the application it serves as well as the technology it uses become more complex. Figure 1.1 depicts this relationship. The 1:1 correlation between FDA Quality System Regulation provisions and business needs is shown in Table 1.1.

Business Growth, Competition, and FDA Regulations

All businesses have to understand the markets or market segments they serve. This is a basic input to develop business growth strategies that are usually shaped by competition. Business strategies are typically based on planned actions aimed at dealing with the forces of competition (Porter 1979), which include:

TABLE 1.1 Relationship Between Business Needs and the Quality System Regulation

Business Need	Quality System Regulation Provision
Understanding and meeting customer requirements	Design input
Meaningful process and product specifications	Design output, design verification, design validation, design transfer
Consistent and reliable measurements for process parameters and product attributes	Test method validation, process validation and process controls, design output, design verification, and design validation
Consistent process	Process validation, design transfer

- the threat of new entrants,

- the bargaining power of suppliers,

- the bargaining power of customers,

- the threat of substitute products, and

- jockeying for position among current competitors.

Design controls are basic tools to differentiate the firm's positioning among competitors, and thus they constitute a strong element of a business strategy. The global opening of the markets promises to transform many medical devices into commodities, if this has not already happened. This will lower the profit margins for most medical device companies, especially those with an average or weak positioning. The logical strategic step to survive will be to cut the cost of manufacturing operations as much as possible. For example, finding a tax haven or a low-wage country for production are already strategies of some companies. Firms with poorly designed manufacturing processes will be facing typical GMP problems (e.g., process deviations, or the process being transferred to the new facility not passing the validation protocol). A tactical plan will call for an added expense known as the technical services group. Unfortunately, many business leaders believe that such tactics are needed only for compliance with FDA regulations.

The logical step to grow the business will be to design "innovative" products. In the finance world, these mean "big profit margin devices." These substitute products are used as the driving business strategy: The claim is made that the new device will make healing faster, or will reduce the days of hospital stay, or will perform in some other "improved" fashion. Traditionally, when the business strategies are prepared, the cash flow analysis, with its many assumptions, is what drives the planning process. Many of these assumptions (e.g., final price, expected overhead, and time to market) become unrealistic goals for the design and development team. If the firm has not adopted design controls as a tool for market differentiation, unrealistic goals can come into conflict with regulations as a desperate design and development team struggles with unclear customer requirements and/or lack of qualified team members and resources to accomplish the business goals and still be in full compliance with the design control and all other regulations.

Why Quality Systems Rather Than Just a Set of Regulations?

Throughout this book, we emphasize the interrelationships among the various elements of regulation because the Quality System Regulation per se is a system. During the design and development phases, the design and development team must have the means to plan, document, execute, and show compliance with customer and regulatory requirements. Therefore, such a team is not exempt from knowing and understanding the firm's quality system (QS). A standard definition for a quality system is "the organizational structure, responsibilities, procedures, processes, and resources for implementing quality management." The output coming from the design and development team must be "embedded" within the firm's QS. Such practice may bring the design and development disciplines needed to control design.

QSIT

QSIT stands for "quality system inspection technique," a new FDA inspectional process that can be used to assess a medical device manufacturer's compliance with the Quality System Regulation. This technique is expected to help the field inspector focus on key elements of a firm's quality system. It approaches a firm's quality system by using the logic of a fault tree analysis (FTA) technique. In other words, it is a top-down approach by which the investigator evaluates the basic elements in each quality subsystem by sampling quality records. "Top-down" also means that the inspection begins by evaluating whether the firm has addressed the basic requirements by defining and documenting appropriate procedures followed by verification of their implementation. QSIT provides for a systematic quality audit in which looking at problems is at the bottom of this approach. The goals of this strategy are to:

- decrease the inspection time,
- increase the focus on the key elements of the firm's quality system,
- move toward harmonization,
- ensure Quality System Regulation coverage,

- increase consistency, and

- improve review efficiency.

These six goals represent a win–win situation for both the FDA and the industry, as demonstrated by results of early studies reported by the CDRH (FDA CDRH 1999c). QSIT guidelines prepared by the FDA divide the Quality System Regulation into seven subsystems, the last four of which are key quality subsystems (FDA Office of Reg. Aff. 1999):

- Facility and equipment controls

- Material controls

- Records/documents/change controls

- Management controls

- Design controls

- Corrective and preventive actions (CAPAs)

- Production and process controls (PAPCs)

It is important to realize the interdependence of these last four quality systems and the fact that CAPAs and PAPCs are supposed to be implemented as part of the design and development work—in particular, during design verification, validation, and transfer.

Examples of Lack of Design Control

The following are typical examples of design control problems that the authors have found during their practice:

- Failure to properly identify and establish adequate physical and performance criteria for the device before production

- Failure to verify that the device met physical and performance requirements before production

- Failure to ensure that device components functioned properly in conjunction with other components

- Failure to ensure that the environment would not adversely affect components

- Failure to select adequate packaging materials

- Failure to validate software prior to routine production

TABLE 1.2   Similarity Between Reliability Engineering Principles and the Quality System Regulation

Classical and Modern Reliability Engineering	Quality System Regulation
Quality by design	Design verification and design validation
Quality of conformance	Design transfer, process validation, process controls, and corrective action

All of these examples are typical considerations taken by reliability engineers during the traditional design and development cycle. Reliability,[2] as a discipline, deals with two general concepts of quality:

- Quality by design

- Quality of conformance

The similarity of these two concepts with the new regulation is clearly shown in Table 1.2. In FDA terms, we may say:

- Quality by design is achieved via design verification and design validation.

- Quality of conformance is achieved via design transfer, verification, validation, and control of the manufacturing process. When deviations occur, quality of conformance is achieved via corrective and preventive action.

An important concept here is that design control actually includes both the product and the manufacturing process. In fact, in the research and development (R&D) or new product development world, manufacturing and process engineers often say, "Whoever designed it shall validate it." This stance signifies, to us, such issues as lack of design for manufacturability and designing with "unrealistic specifications."

Why Control the Design of Products—Aren't the Traditional cGMPs Enough?

The inherent or intrinsic quality, reliability, durability, manufacturability, testability, stability, effectiveness, and safety of a product are defined in the design and development stages, based on customer needs and usage as well as on the applicable environmental limitations and circumstances. In fact, this is a typical set of design inputs. This is a change of mind-set for the medical device industry, which has become accustomed to inspecting products during manufacturing and sometimes expending a significant

---

2. As a formal science, this is known as Reliability Engineering.

amount of company resources on "process and product babysitting." Consider the following possibilities:

- The manufacturing process is perfect, but the design is faulty.

- The design may have all the necessary attributes, but it is not manufacturable (e.g., it is an inconsistent or nonreproducible process).

- The design is perfect, but it does not meet the intended use of the customer; or, if it does, it harms the patient, the user, or the installer.

The goal of the Quality System Regulation is to achieve safety and efficacy by focusing on eliminating these possibilities.

Why Good Documentation, Quality Assurance, or Quality Science Is Not Enough

Design control and even process validation will present a challenge to the traditional Quality Assurance (QA) manager, who has historically been monitoring engineering and manufacturing operations for deviations and nonconformances. The documentation system does not confer quality on the product. If the documentation presents instructions that do not produce a device that meets the intended use as well as the reliability requirements, then the QA role is merely paperwork, no matter how well the firm follows its documentation system. The new role of the quality function in the medical device industry is to provide technical and scientific depth. Design controls provide a firm's management or decision makers with the visibility of the design and development process. Thus, knowledge-based decisions can be made.

Changes in Technical Education and Training

Engineering and life sciences[3] education is basically deterministic; the students are not sufficiently exposed to the concept of variability. Usually, not until they enter the world of industry do they first see concepts about statistical process control (SPC) and error of measurement (e.g., Gage Repeatability and Reproducibility [GR&R]). In fact, our experience as industry trainers and consultants is that even the basic concepts of hypothesis testing and confidence intervals are not correctly understood; and furthermore, they are not well applied.[4] Therefore, medical device firms should overcome this inherent shortcoming by incorporating tools that manage and consider

---

3. These are the two major broad areas of education involved in the design and manufacture of medical devices and pharmaceutical products.

4. This is true, even for personnel with doctorates.

variability into their design and development cycle. Even some "veteran" design experts rely completely on the early results of a single prototype (without considering statistical concepts such as variability, confidence, and probability) and make an immature decision to order the design to be released to the next (development) phase. This is a vivid example of what a lack of discipline in the design stage really means and what the consequences of such actions can bring during the product life cycle. Similarly, design inputs coming from marketing specialists can be shifted from interview to interview if a population and sampling scheme is not well defined.

Summary

This chapter establishes a clear link between design control guidelines and expected business outcomes. By complying with design control regulations, any medical device firm can bring its innovations to reality in a systematic manner, just like a high-tech company. This will enable such firms to become more competitive. In the next chapter, we discuss in detail the elements of the design control requirements.

CHAPTER TWO

Design Control Requirements

Design control, as a requirement by the FDA, became part of the new Quality System Regulation placed in effect on June 1, 1997. The 12-month time span from June 1, 1997, to June 1, 1998, was called the "transition period." During that period, design control was not used as an enforcement tool if the manufacturers could show they were taking "reasonable steps" to come into compliance. The actual first step many manufacturers took was to understand the regulation and prepare themselves for the development and the adoption of new quality systems. This step was a major change to many medical device manufacturers. Design control became an enforcement tool on June 1, 1998.

The regulation on design control applies to any class III or class II device and certain class I devices. Design controls must be used with any class I device with software as well as those listed in Table 2.1.

Table 2.2 depicts the design control requirements and some typically associated quality systems. These quality systems can be viewed as one element of the firm's mechanism to comply with the regulation.

TABLE 2.1   Class I Devices Subject to Design Controls

CFR Section	Device
868.6810	Catheter, tracheobronchial suction
878.4460	Glove, surgeon's
880.6760	Restraint, protective
892.5650	System, applicator, radionuclide, manual
892.5740	Source, radionuclide teletherapy

TABLE 2.2 Design Control Requirements (21 CFR 820.30) and Associated Quality Systems

Requirement	Typical Associated Quality Systems
a. General	Preparation of quality policies and procedures.
b. Design and develop-ment planning	Specific procedures for product design and development planning and design change planning.
c. Design input	Procedures for data collection, analysis, and storage (filing) of customers', users', or installers' historical field quality data. Examples include focus groups with doctors or other healthcare givers, panel discussions, interviews, surveys, field complaints, MDRs, and human factors engineering (e.g., ergonomics, industrial design). How to execute, document, analyze, and store such information.
d. Design output	Procedures for translating design input into engineering or scientific design specifications. How to execute, document, analyze, and store such information. Procedures for planning, executing, and documenting experimental protocols, such as design verification and validation.
e. Design review	Procedures for organizing, executing, and documenting design reviews. Procedures for defining a design and development team roster and reviewers. How to document pending issues and how to follow up and close all of them. How to execute, document, analyze, and store such information.
f. Design verification	Procedures for software/hardware verification. How to execute, document, analyze, and store such information.
g. Design validation	Procedures for software/hardware validation of animal studies, clinical studies, and cadaver laboratories. How to execute, document, analyze, and store such information.
h. Design transfer	Procedures for the preparation of DMR, process validation (IQ/OQ/PQ), and training.[a] Supplier or contract manufacturer certification. How to execute, document, analyze, and store such information.
i. Design changes	Procedures for changes and updates to "preproduction."[b] How to execute, document, analyze, and store such information.
j. Design History File (DHF)	Procedures for creating, approving, and updating the DHF. How to execute, document, analyze, and store such information.

[a]DMR = Device Master Record; IQ/OQ/PQ = installation, operational, and performance qualifications.

[b]While the design and development process goes through its iterations, temporary DMR elements (e.g., drawings, specs, test methods) are generated. Some firms define the approval of such DMR elements as "conditional release" and/or "non-salable release." This "preproduction" change control system is different from a change control system after the product has been approved or released to the marketplace.

What Is Design Control?

A practical definition for design control can be: "a set of disciplines, practices, and procedures incorporated into the design and development process of medical devices and their associated manufacturing processes." The definition from the Global Harmonization Task Force (GHTF) is similar, but lacks the word "disciplines." A product development team without appropriate development disciplines may adopt the design control requirements merely as more "paperwork." Company management and R&D team leaders or managers must accept this painful reality. This idea is reinforced by the historical tendency for R&D teams to be driven and rewarded only by "time to market." Design control as a discipline has been with us since the 1960s. Most of the information contained in the regulation and the FDA guidelines is an application of traditional reliability management programs and former military standards. Thus, three key definitions must be understood and distinguished from each other:

- Discipline, defined as a system of rules governing activities

- Practice, defined as doing something customarily

- Procedure, defined as a series of steps followed in a regular definite order

R&D teams have their "way of doing things" (i.e., practice). Unless the adoption of design controls is strategized, these teams merely see more procedures to follow with no practical value to be added. The main challenge for R&D managers and R&D–quality managers is to bring design controls to the level of a discipline while attaining company financial and regulatory goals.

What Design Control Is Not

The FDA design control regulation is not a detailed prescription for the design and development of medical devices. Furthermore, it does not challenge the science, the development modus operandi, the "inventive stages," or the engineering knowledge in product design and development. FDA investigators will evaluate the process, the methods, and the procedures for design control that a manufacturer has established. Then they will evaluate compliance to such methods and procedures.

Technical Considerations: The Basics of Product Development

Basic considerations should be taken during the process of designing a medical device. It is easier to think in terms of past product recalls and/or medical device reports in order to generate the following list of special challenges:

- Materials selection (e.g., effect from sterilization technology, biocompatibility, toxicity)

- Process changes (e.g., strength decay on a plastic part upon "process improvements" to an injection molding process in which the mold temperature is set at a lower temperature to increase throughput[1])

- Product characteristics such as:
  - Chemical (e.g., the toxic plasticizer DEHP in polyvinyl chloride [PVC] tubing)
  - Electrical (e.g., the potential effect of static on a diagnostic electronic analyzer)
  - Magnetic (e.g., field effect and control, electromagnetic interference/compatibility [EMI/EMC])
  - Physical (e.g., ergonomics, human factors considerations)
  - Biological (e.g., light sensitivity in some reagents of in-vitro diagnostics)

- Energy modality instruments and their potential energy transfer to human beings (e.g., leakage current from radio frequency [RF] device)

- Misuse of a device (e.g., electrosurgery in a finger causing thrombosis, or starting a fire in the trachea)

- Abuse of a device (e.g., using in-vitro diagnostics reagents, calibrators, or controls beyond the expiration date, and/or not calibrating the analyzers as suggested by the manufacturer)

- Hazards in the absence of failure (e.g., leaving an energized electrosurgical electrode on the skin of a patient)

Although basic in nature, some of these considerations have been stated as root causes for product recalls, field complaints, and other signals of inappropriate and/or suboptimal performance or safety levels. These challenges can be avoided by adoption of such appropriate design control disciplines as risk analysis.

Scope of Application

Design control is not intended[2] to be applied to the following stages of the design life:

- Technology discovery and assessment

- Concept feasibility

---

1. The authors have seen such an effect on some plastic components; however, this is just an example of a process change and should not be interpreted as a default situation.

2. Although, many of the tools and the discipline of design control can help the research team to achieve concept feasibility at a faster pace.

Thus, design control applies once the feasibility of the concept and technology have been proven and approval has been obtained to develop the product. In some large companies, this is known when management initiates a "capital request" type of action. However, we recommend that device companies define the action that triggers formal design control in their design control policy, in the quality manual, or in the design control Standard Operating Procedure (SOP).

Practical Implementation of a Design Control System

Some of the steps necessary to implement a design control system, in chronological order, are:

1. Define the design and development process of the firm—for example, technology development and discovery, concept development and feasibility, design works, prototyping, testing, pilot runs, and reviews. The firm defines where design controls will actually start to apply.

2. Develop policies, procedures (see Table 2.2), and work instructions for appropriate control of the design and development process of the device and its manufacturing process.

3. Develop policies, procedures, and work instructions for risk analysis.

4. Develop a training plan. Typical skills that medical device companies need to strengthen include quality systems for "non-quality personnel," compliance with the regulation, reliability engineering, use of external standards, six sigma, FMEA, FTA, and statistical methods for nonstatisticians. Specifically, the training plan should address how to create consciousness within the design and development organization and develop a discipline for design excellence.

5. Define internal interfaces and roles. For example, if a new manufacturing process has to be developed, the product development team will need to interface with a manufacturing process development engineer or specialist.

6. Review quality systems for adequacy. For example, a company that has been manufacturing plastic and metal-based devices is moving toward designing capital equipment with electronic components, or vice versa.

The preceding steps are also elements of the necessary transformation, as the following three examples show:

- Traditionally, quality engineers (QEs) have worked on purely manufacturing applications as well as process validation and the documentation needed for product or technology transfer. Now, design control opens the door of R&D to QEs just as in electronics and other "high-tech" industries. There will be a need and an opportunity for QEs with strong reliability backgrounds to have a more proactive role. Design and development teams will be in need for such human resources to bring the disciplines and tools, such as quality function deployment (QFD), FMEA, and risk analysis, needed for design control.

- Technical interfaces expand the concept of the R&D team beyond the conventional "design and development team" and formally assign roles and responsibilities to "technical and logistical interfaces." Thus, design control applies to departments other than R&D and Quality. Training on design control and awareness is also essential to the other areas of the company.

- The documentation system is essential for design control, and a Design History File should be generated. The challenge lies in how to bring such a level of discipline and organization to R&D engineers and scientists. Also, depending on what the current and new products are, additional needs for quality systems may exist. For example, a maker of disposable diagnostics products (e.g., home testing for pregnancy) is now going to produce sophisticated random access electronic analyzers and their respective reagent kits. This immediately prompts the need for systems to evaluate and control the life of capital equipment, and software reliability, among many other considerations.

Design Controls and Investigational Device Exemptions

Originally, devices being evaluated under **investigational device exemptions** (IDE) were exempted from the original GMP regulation. This makes sense for two reasons: (1) such devices were mostly "lab made," with no production equipment or no mass manufacturing processes, and GMPs were not really applied at that time, even though sponsors were required to ensure manufacturing process control; and (2) such devices might never be approved for commercial distribution. However, starting with the preamble to the current regulation, the FDA has stated that "it is reasonable to expect manufacturers who design devices to develop the designs in conformance with design control requirements and that adhering to such requirements is necessary to adequately protect the public from potentially harmful devices." The FDA sees design control requirements as "basic controls needed to ensure that the device being designed will perform as intended when produced for commercial distribution."

Our past experience in the electronics, automotive, and telecommunications industries tells us that even for investigation purposes, it is sound business and economic strategy to develop and build experimental models or prototypes under design controls. If the investigation fails, the design and development or research team usually depends on the amount and quality of information gathered to find a root cause for the problem. Design FMEAs (DFMEAs) generated during this early phase are precisely aimed at helping the designers to systematically uncover issues. Conversely, if the investigation is a success, part of the design outputs are already defined, and their justification might be readily available. Thus, the DHF is partly built and the clinical evaluations can be part of the design validation.

Another consideration is that in an IDE evaluation, human beings are being exposed to the potential hazards of the device under investigation. By following design control guidelines, the IDE device will have a risk analysis that can also avoid delays and injuries by proactively mitigating the identified hazards. Results from an IDE usually motivate design changes. By applying the disciplines of design control, the firm can also ensure controlled design changes and better evaluation of results.

Our view is that the FDA is once again providing for a win/win situation. The improved research and development disciplines should minimize public and financial risks.

Design Control Requirements

This section introduces and discusses each design control requirement. It provides practical examples as well as tables that relate each 21 CFR 820 requirement with typical quality systems audit questions for the purpose of helping firms to execute their own self-assessments (Tables 2.3 and 2.4).

General Requirements

The intent behind applying design control guidelines to the product development process from the completion of the concept and feasibility stage to the full-blown production stage is to ensure that the product design meets the specified requirements and that it can be reliably produced.[3] This can be accomplished by establishing, utilizing, and maintaining effective quality system procedures that monitor and verify all design works at all stages of the design process. The regulation is very general and flexible. Each firm has the opportunity to develop its quality systems procedures to comply with the regulation in a manner compatible with its own products and technologies.

Design and Development Planning

Plans should be produced that allocate the responsibility for each design and development activity. Somehow, the design and development process has to be controlled, and the product development team or R&D management should have a sense of where in

---

3.   The regulation establishes the foundation for design transfer (i.e., process validation). As discussed in Chapter 3, the real essence of process validation is reproducibility and repeatability.

TABLE 2.3 General Requirements for Design Controls

21 CFR	Key Questions	Your Company's Response (Yes/No/Maybe)
820.30(a)	Does your company have procedures to control and verify the design and development of your products? Do they cover all 10 of the design control requirements?	
	Are these procedures used to uniformly guide the design and development process?	
	Are all design control elements covered by the firm's procedures?	
	Do you monitor if these procedures ensure that all the specified requirements are met?	
	Are the product designers, managers, and/or scientists actually complying with the procedures?	

this process they are at all times. Each of these activities should be referenced within the plan. This should be an ongoing process until the design is completed, verified, and validated. Whoever is in charge of generating the design and development plan (DADP) should keep in mind that the underlying purpose is to control the design process, which is aimed at meeting the device's intended use and its associated quality objectives.

The benefits associated with a design and development plan include the following:

- A disciplined approach to project management. Thus, knowledge-based decision making becomes plausible.

- A project-specific plan (it includes specific details).

TABLE 2.4 Design and Development Planning

21 CFR	Key Questions	Your Company's Response (Yes/No/Maybe)
820.30(b)	Do plans identify the responsibility for each design and development activity? Are all activities included?	
	Do plans define the technical interfaces?	
	Do plans describe or reference the activities? Are they updated as the design evolves?	
	Are qualified personnel equipped with adequate resources for the planned activities? Have you included your outsourcing partners, supply chain, and project consultants?	

- A common communication mechanism ("everybody is on the same page"). It is a project road map.

- Proactive planning (no surprises).

- Inclusiveness. Regulatory, marketing, economic (e.g., cost of goods sold [COGS]), and quality requirements are included in one structure, which facilitates alignment for all parties involved or with a stake in the project. This is the chance to bring the organization together and adopt the new terminology (e.g., DMR, DHF, design validation).

- Ease of project issue resolution.

- Overall compliance record and traceability (why we did it like that).

In practical terms, the DADP, once approved, really becomes a checklist of activities to be performed (although this was not intended) as well as a management control tool. Typical project management tools, such as Gantt charts, the program evaluation and review technique (PERT), and the critical path method (CPM) can be used here.

Some of the typical questions that a plan should answer include the following:

- Do we have a complete team (e.g., is the team missing a specialist in software reliability)?

- Do we have independent reviewers? Who are they? What are their roles or areas of expertise?

- Have we established all the interfaces? Do these include supply chain and consultants?

- Does everybody know what is expected from them? Do they know their due dates and deliverables?

- When is the team supposed to make what decisions?

- Are we going to meet the business targets (e.g., schedule, cost, risk, and reliability levels)?

- Where are we on this project?

- Where, when, and how can top management help?

- What is the critical path? How can we ensure that falling behind schedule will not compromise quality and reliability?

- What is the goal for manufacturability?
  - □ How do we measure manufacturability?
  - □ What do we mean by manufacturability (e.g., Cpk or Ppk[4] > 1.33, COGS < 25%, etc.)?
  - □ Are we missing any quality system requirements, such as biocompatibility, risk analysis (RA), system FMEA (SFMEA), design FMEA (DFMEA), process FMEA (PFMEA), environmental impacts, component qualifications, accelerated aging, pre- and poststerilization correlation analysis, stability of process, stability of product (shelf life), stability of packaging, validation of test methods, GR&Rs, measurement capability analysis, and so on?

Make sure that a procedure for design and development planning is not finished until a full and thorough assessment of any quality systems gaps is completed. Modeling the business and its standard jargon can be useful in identifying the quality systems gaps (e.g., a firm that has always done cobalt sterilization now has a project in which ethylene oxide is the only sterilization alternative[5]). For each project, the company procedure for DADP should seek answers to the following general questions:

- What is to be designed and developed (e.g., what are the goals and objectives)?

- What are the design and development activities to be executed? In what order[6] should they occur? What is the purpose of each activity? How does each design and development activity comply with the regulation?

- What are the milestones and deliverables at each design and development phase? How do the design and development teams know when to proceed? How is that documented?

- Which quality systems, procedures, and work instructions will be used?

---

4. Cpk is typically defined as the process potential capability ratio. Ppk is typically defined as the process performance ratio.

5. Thus, where will they sterilize? What company procedures will be used to validate the sterilization process? If an outside contractor (e.g., purchasing controls) will be used, then this new supplier may need certification.

6. An auditor of quality systems may expect the manufacturer to paint a clear picture of the flow of the design process in order to establish where the "predevelopment" activities (e.g., concept or technology development and design feasibility) ended and where the actual design and development work starts. Also, ideally, the DADP is expected to describe the phases of design (e.g., preliminary design, early prototype, evaluation, design review I, redesign, prototype, testing, evaluation, verification, pilot runs, design review II, final adjustments, final verification, complete DMR [prior to design transfer], transfer to manufacturing, validation of process and design, complete final DMR [after design transfer], final design review, complete DHR).

- What qualifications[7] are needed for those who will perform the design and development tasks? What qualifications are needed for the reviewers?

- Which standards and other regulations are applicable (e.g., Policy 65 in California)?

- Who is responsible for what and when is the due date?

- What are the required levels of quality and reliability?[8]

- When is the design "frozen" (i.e., no more changes beyond this point)?

- How is the DADP reviewed,[9] updated, and approved as design and development evolves?

- How is project progress measured?

- Where and when will the design and development team face critical decision points? For example, the preliminary hazard analysis may lead to a failure mode identified in the DFMEA that indicates that a major design change is needed to mitigate a hazard. Typically, upper management assumes that once there is a DADP, everything will happen. Sometimes, the design and development team does not know what they do not know!

- How does the DADP map into the design and development phases? Table 2.5 provides an example.

Advice to Management

Management should heed the following advice:

- Avoid using the plan merely as a checklist.

- Rather than questioning the design and development team regarding task completion, question the team regarding achievement of safety, risk, and performance levels. A better question might be: "Are we meeting the quality and reliability goals?"

---

7. This is an example of the interrelationship between design control and the training quality system. Are your training records up-to-date? Do you have evidence of your personnel's credentials?

8. Usually, a complementary plan of quality and reliability accompanies the DADP. Thus, such quality and reliability goals may not necessarily belong in the DADP.

9. A bit of advice to design review and design team leaders is to avoid the temptation of reviewing status only. The value of the design review primarily resides in the prompt identification of issues affecting safety and performance. In practice, most of the attention is given to the schedule, which is not the intention of the regulation.

TABLE 2.5   Design Control Requirements and the Design and Development Phases

Requirement	Design and Development Phase[a]
Risk analysis	Started at preliminary design (essential part of design input). Closed upon successful validation.
Design input	Preliminary design.
Design output	Started at preliminary design. Reviewed upon successful verification. Closed upon successful validation.
Design review	After evaluation of preliminary prototype. After pilot runs. Final review upon full validation of process and design.
Design verification	Preliminary, upon prototype testing and its evaluation. Final upon final adjustments to the design.
Design validation	Upon successful final verification and completion of the DMR and the DHF.
Design transfer	Started upon completion of the DMR and prior to validation. Completed upon final approval of the complete project.[b]
Design changes	May occur throughout the entire project. The key elements are its management, control, and documentation.
Interfaces	First defined at preliminary design. Will change as the project evolves. For example, a market research analyst may be needed early in the project, but not during design transfer. Just the opposite would be true with the factory materials manager.

[a]These phases are merely examples. Each firm defines its own phases for design and development and associates such phases with the design control requirements.

[b]Project completion may require activities beyond the design control regulations (e.g., final cost analysis and negotiations, submitting a "Technical File" for CE [Conformité Européen] marking). Thus, we are splitting design transfer into regulatory and business needs.

Technical Interfaces

Several groups of personnel may provide input to the design process. Any organizational and technical interfaces among these groups must be clearly defined and documented. This information should be reviewed regularly and made available to all groups concerned. Technical interfaces are an interdependent part of 820.30(b), Design and Development Planning, as shown in Table 2.6.

Examples of technical interfaces include the following:

- Packaging and sterilization engineering design and development

- Technology development or advanced research group

- Material sciences

- Clinical, medical, veterinary, and regulatory affairs

- Customer support

TABLE 2.6   Technical Interfaces

21 CFR	Key Questions	Your Company's Response (Yes/No/Maybe)
820.30(b)	Are the organizational and technical interfaces between different groups defined and documented?	
	Is information regarding the various interfaces within the organization regularly reviewed?	
	Is interfacing information transmitted to all concerned groups within the organization?	

- Toxicology sciences

- Patent engineering

- Graphics design (e.g., labels, inserts, user manuals)

- Field quality (e.g., complaint handling and analysis, MDRs)

- Materials management and/or production planning

- Supplier management

- Supplier quality assurance

- Suppliers (e.g., raw materials, components, subassemblies, contract manufacturing)[10]

- Purchasing

- Facilities and utilities engineering

- Marketing

- Consultants

- Business systems

- Manufacturing engineering

- Quality assurance/quality control/testing services or independent laboratories

- Reliability engineering

---

10. Outsourcing seems to be "the strategy" today. When the final assembly of a product is made by an external contractor, that contractor shall also participate in the preparation and review of the design and development plan. After all, they will execute the main activities of design transfer. In other instances, they might be doing part of the product and process development.

TABLE 2.7 Example of a Project's Roster[a]

Examples of Typical Design and Development Tasks	Mrktg.	R/A	Clinical	Rel. Eng.	Mfg.	R&D
Product Requirements	P			D		S
Design Specification	S			S		P
Regulatory Strategy	S	P		S		S
First Design Review	P	P, D	P, D	P, D	P	P

[a]P = primary milestone or deliverable responsible; S = secondary or support function; D = design reviewer; Mrktg. = marketing; R/A = regulatory affairs; Rel. Eng. = reliablity engineering; Mfg. = manufacturing.

- Information management systems or Information Technology (IT) groups

- Certifying bodies (e.g., the Canadian Standards Association [CSA] and Underwriters Laboratories [UL])

- Warranty services

- Industrial design (e.g., human factors engineering, ergonomics)

Company Strategies

Based on the qualification of human resources and their functions, a design and development matrix can be used in the DADP to assign responsibilities within the design and development team as well as the technical interfaces and design reviewers. The matrix in Table 2.7 is an example. In fact, this matrix can also define deliverables prior to a design review.

Design Input

Design input can be defined as performance, safety, business economics, and regulatory requirements that are used as a basis for device design. The purpose and intended use of all medical devices should be clearly understood so the design inputs can be identified and documented. The company should review these inputs and any inquiries should be resolved with those responsible for the original specification. The results of contract reviews should be considered. Table 2.8 specifies key questions with regard to design input.

Design input presents in various ways. The two examples in Table 2.9 illustrate that when we hear the customer, we are not going to get "direct usable" design inputs. Such information has to be interpreted and massaged to be able to specify design requirements. For example, you can never expect a customer to tell you what kind of plastic resin you have to use to meet some of his or her needs for a medical device.

TABLE 2.8 Design Input

21 CFR	Key Questions	Your Company's Response (Yes/No/Maybe)
820.30(c)	Are product design input requirements identified and documented?	
	Are the design input requirements selections reviewed for adequacy?	
	Are incomplete, ambiguous, or conflicting requirements resolved with those responsible?	
	Are the results of contract review activities considered at the design input stage?	
	Are statutory and regulatory requirements considered at the design input stage? What about standards?[a]	

[a]For example, a key design input when designing electrical medical devices is compliance with the IEC 60601-1 standard.

TABLE 2.9 Examples of How to Go from Raw Design Input into Design Requirements

	Customer Requirements	System Requirements	Design Input
Practical Interpretation	External Customer Needs and Internal Goals	Measurable Customer Requirements	Design Requirements
Example 1	... can be used on big and small humans ...	Targeted at 90% of domestic potential patients	Based on small, medium, and large sizes
Example 2	... most reliable device in its class ...	... Total reliability = 99.7%	Reliability allocation for three main subsystems = 99.9%[a]

[a]By using the simple principles of probability in systems reliability, we can allocate the same reliability of 99.9%, (thus .999 × .999 × .999 = 99.7%) to the three main systems.

In assessing your design input procedures, typical audit questions include the following:

- What mechanism(s) are used for dealing with unclear or conflicting requirements (e.g., QFD, FMEA, risk analysis)?

- At what stage of design and development were the design inputs reviewed and approved?

- Are the design requirements appropriate (i.e., is the intended use being met with those requirements)?

- Will it be safe?

- Are human factors being considered as part of design inputs?

Human Factors

The Quality System Regulation requires manufacturers to address human factor issues or considerations during the design and development phases. An obvious consideration is the interface between the device and the patient and/or healthcare provider. However, a great deal of variability exists among patients and healthcare givers. Such sources of variance range from physical to sensory to cognitive abilities. As an example, consider a case in which intravenous solutions are being administered to a patient by means of a computerized drug delivery system with three peristaltic pumps. When one of the medical solutions being administered is finished, the nurse is supposed to stop the equipment. But if she got confused with the control menu of the system, she would end up performing all the operations manually.

Human factors issues are related not only to patients and healthcare givers, but also to installers, biomedical engineers, and technicians who provide for maintenance and repair.

Design input regarding human factors should consider the different environments in which the device would be used. For example, will the device be used in trauma rooms or just during doctors' visits. A very important consideration is intuitive design. Many healthcare givers openly admit to not having the time to read manuals or instructions for use.

Human factors considerations can eventually define reliability goals. For example, in many cases the healthcare professional lifts a footswitch for electrosurgery by pulling the cable that connects to the generator. On other occasions, the footswitch is dropped to the floor during a quick connection action. The new FDA guideline document entitled "Medical Device Use Safety: Incorporating Human Factors Engineering into Risk Management" provides more emphasis on human factors considerations in design to minimize risks (FDA 2000).

Examples of Conflicting Requirements

The following scenarios exemplify the conflicting requirements that can occur in design input:

- Diagnostics reagent company ABC wants to design an ultrasensitive assay that also covers "above normal high" concentration values of a given hormone (e.g., the dynamic range is too wide, thus, sensitivity may be compromised).

- Company XY wants to develop a surgical device based on requirements provided by right-handed surgeons that are in conflict with requirements provided by left-handed surgeons.

■ A survey conducted for a new medical device concept indicated that 50 percent of the surgeons wanted a tactile feedback mechanism in the product, and the other 50 percent did not want it because they thought that the mechanism was an indication of a defect.

Some Myths About Design Input

The following myths about design inputs are illustrative of misconceptions surrounding this important design factor.

■ *Design inputs are to be gathered only by marketing.* Although marketing research plays a very important role in gathering design inputs, not all the relevant questions and data gathering will come from "marketeers." Design inputs occur in very different ways, as seen in the examples depicted in Table 2.10.

■ *In addition to marketing, only design engineers are to gather design input data.* Table 2.10 shows that design inputs can be related to standards engineering, reliability engineering, industrial design, business development, regulatory affairs, and so on. The entire design and development team and even some of the technical interfaces should participate in the design input activities. For example, as design and development reliability engineers, we have asked surgeons and surgeon assistants to define strength requirements and user conditions for surgical devices. These crucial design inputs define reliability design goals.[11]

■ *Design input has nothing to do with reliability.* By definition, this is a big myth. Implicit in the definition of reliability are the following design requirements:

  □ Reliability is quantified in terms of probability.

  □ Function or intended use must be defined.

  □ Environment or operating conditions must be defined.

  □ There is an operating time between failures.

■ *Design inputs are general requirements.* The effectiveness of the entire design and development program starts with the quality of details stated for design inputs. Depending on the nature of the input, some details may have to wait for the design and development team to define the design outputs.

Sources of Design Inputs

Design inputs can come from a variety of sources, such as the following:

■ Surveys with doctors, clinicians, physician assistants, and patients as direct users of medical devices

---

11. Marketing and design engineering personnel typically never look for these types of questions and information.

TABLE 2.10   Design Input Examples

Potential Design Input	Examples
Intended use	Specific versus general surgery instrumentation.
	Endoscopic or open surgery?
	Screening or final determination abused drug immunoassay?
	Beating or still heart surgery?
User(s)[a]	Installer, maintenance technician, trainer, nurse, physician, clinician, or patient?
	What is the potential user's current familiarity with this technology?
Performance requirements	Highly sensitive immunoassay or with a very broad dynamic range?
	How long will the surgical procedure last?[b]
	Is there a potential complication with very big, very small, obese, or skinny patients?
	Frequency of calibration longer than a month for a diagnostic assay?
	Software user interface requirements.
	Software requirement specifications.
Chemical characteristics	Biodegradable?
Compatibility with user(s)	Biocompatibility and toxicity.
Sterility	Pyrogen free?
	Sterile?
Compatibility with accessories/ ancillary equipment	IV bag spike or other standard connectors?
	Electrical power (e.g., US vs. South America)?
	Open architecture for computer systems networking.
Labeling/packaging	Languages, special conditions?
	Heat protection?
	Vibration protection?
	Fragility level?
Shipping and storage conditions	Bulk shipments or final package?
	Humidity and temperature ranges?
Ergonomics and human factors	International versus domestic considerations.
	"Foolproof"?
Physical facilities dimensions	Cables for electrosurgical generators in Europe may need to be longer in European than in US operating rooms.
	The same would apply to devices that include tubing for blowing CO_2; for example, a blower with mist for a coronary artery bypass graft (CABG) that is used to clean the arteriotomy area from blood. The length of the tubing has to be longer for Europe than for US operating rooms.
Device disposition	Disposable versus reusable.
Safety requirements	UL/IEC/AAMI requirements[c].
Electromagnetic compatibility and other electrical considera- tions	Electrostatic discharge (ESD) protection.
	Surge protection.
	EMI/EMC meet IEC standards for immunity and/or susceptibility.
Limits and tolerances	Maximum allowable leakage current on an electronic device.
Potential hazards to mitigate	Potential misuses such as warnings and/or contraindications in inserts or user manuals.
	Hazards in absence of a device failure (e.g., electrocution of an infant with metallic probes of a device).

continued next page

TABLE 2.10   *continued*

Potential Design Input	Examples
Compatibility with the environment of intended use	A metallic surgical device that may contact an energy-based device during surgery could conduct energy, thus potentially harming the other organs of the patient.
Reliability requirements	0.99 at 95% at the maximum usage time or conditions. Mean time between failures? Mean time to failure? Mean time to repair? Mean time to maintenance?
User(s) required training	Simplify new surgical instrument and new procedure because it may require complicated training. Programming a handheld blood sugar analyzer.
MDRs/complaints/failures and other historical data	Benchmark from similar, platform, or surrogate devices.
Statutory and regulatory requirements	Policy 65 (California).
Physical characteristics	Dark color in an endosurgical or laparoscopic instrument avoids reflection of light from an endoscope. Amber or dark color bottles are used for filling of light-sensitive reagents.
Voluntary standards	IEEE for electrical components and/or software development and validation.[d] NCCLS for in-vitro diagnostic[e]
Manufacturing processes	Design device such that no new capital equipment is required for manufacturing.

[a]This design input can directly identify a design output such as training requirements. Training is not only for users; business training for sales and marketing personnel is also included.

[b]This is especially important to define the use environment that eventually defines the required reliability. This is an example of the type of question that the R&D quality and/or R&D reliability engineer should be asking during the gathering of design inputs.

[c]UL = Underwriters Laboratories; IEC = International Electrotechnical Commission; AAMI = Association for the Advancement of Medical Instrumentation

[d]IEEE = Institute of Electrical and Electronic Engineers

[e]NCCLS = National Committee of Clinical Laboratory Standards

- Surveys with biomedical engineers and technical personnel involved in the maintenance, installation, and repair of medical devices

- Panel meetings with subject matter experts

- Literature searches and clinical evaluations

- University research reports

- Previous projects

- Industry other than medical device industry

- Hazard analysis on similar devices

- Animal and cadaver laboratories using prototypes

- Database research (e.g., MDRs, field complaints)

- Standards searches

Design Output

The objectives of any new product design should be defined as design outputs. These should be clearly understood and documented. They should be quantified and defined and expressed in terms of analyses and characteristics. Key questions concerning design outputs are delineated in Table 2.11.

Examples of Design Output

Design outputs can vary widely, as the following list illustrates:

1. The device itself

2. Labeling for the device, its accessories, and shipping container(s)

3. Insert, user manual, and/or service manual[12]

4. Testing specifications and drawings (detailed, measurable)

5. Manufacturing (materials and production) and QA specifications or acceptance criteria

6. Specific procedures (e.g., manufacturing equipment installation, work instructions, bill of materials [BOM], sterilization procedures)

7. Packaging feasibility studies, validation testing and results

8. Risk analysis summary, FMEAs, reliability planning and results

9. Biocompatibility and toxicity results

10. Software source code

11. Software hazard analysis

12. Software architecture

13. Software verification and validation (V&V)

---

12. A service manual usually contains instructions for repairs and preventive maintenance. It mainly applies to such capital equipment as that used for magnetic resonance imaging (MRI) or computerized tomography (CT) scans, electrosurgical generators, and diagnostic analyzers.

TABLE 2.11   Design Output

21 CFR	Key Questions	Your Company's Response (Yes/No/Maybe)
820.30(d)	Are procedures in place to ensure that design output meets design input requirements?	
	Are design outputs documented and expressed in terms of requirements and analysis?	
	Do procedures ensure that design output contains or references acceptance criteria?	
	Does design output conform to appropriate regulatory requirements?	
	Are the characteristics crucial to the safe and proper functioning of the product identified?[a]	
	Does design output include a review of design output documents prior to their release?	

[a]Therefore, risk analysis and management is a key design output. As part of your quality systems assessment, a risk analysis and management procedure is a must.

14. Technical file or design dossier for *CE* marking

15. Clinical evaluation results

16. Transit, storage, and shipping conditions testing and results

17. Supplier and component qualification (e.g., the DHF should include evidence of official communication to component suppliers stating status of qualification approval and process control agreements[13]).

In general, the design output deliverables will reside in the DHF and the DMR.

Relationship Among Design Input, Design Output, DHF, and DMR

Table 2.12 is an example of the fact that design output meets design input. Figure 2.1 shows that design output is really an answer to a request (design input) plus the evidence to support the decision. All of the preceding list of design outputs belong in the DHF at any given time, as depicted in Figure 2.2. However, only items 1 to 6 would end up being part of the DMR. The DHF can be seen as a file with re-

---

13. This is another example for the need of appropriate quality systems. A procedure should exist to evaluate and qualify suppliers, which should also describe what documentation is required to notify the supplier when qualification has been attained.

Table 2.12   Example of Design Output Meeting Design Input

Design Input	Design Output	
	Design Specification	DMR
The medical device will be used in trauma rooms. It must be capable of withstanding adverse conditions (e.g., accidental pulling by the tubing).	The bond strength between a luer lock and tubing (IV line) should withstand p pounds of axial force without detaching from the tubing.	The raw material for the luer lock will be X and the solvent Y. Before inserting the tubing into the luer lock, the solvent will be applied, and a curing of T minutes will be allowed.

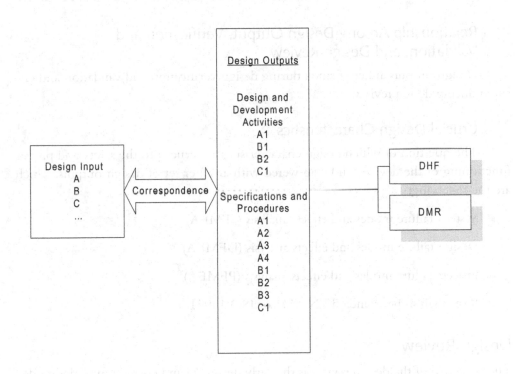

FIGURE 2.1   Relationships among design inputs, design outputs, DHF, and DMR.

cords showing the relationship between design input and design output. The key word here is "records." The DMR is composed of the instructions and criteria needed to make the product. The DHF contains records, whereas the DMR contains "living documents."

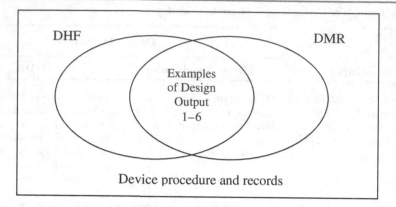

FIGURE 2.2 Some DHF elements will become DMR elements.

Relationship Among Design Output, Verification and Validation, and Design Review

Design outputs are confirmed during design verification and validation and ensured during design review.

Crucial Design Characteristics

The question of which design characteristics are crucial to the safety and proper functioning of the device can be answered with another set of design outputs, which are the following:

- System failure modes and effects analysis (SFMEA)

- Design failure modes and effects analysis (DFMEA)

- Process failure modes and effects analysis (PFMEA)

- Risk analysis per standard EN 1441 or ISO 14971-1

Design Review

The objective of the design review is the early detection and correction of design deficiencies. Such deficiencies can be identified in the different phases of the design and development program. For example, the team can expect more emphasis on design input during the first design review than during the last review. Hazard analysis and the DFMEA are great tools for evaluation, analysis, and discussion among the members of the design and development team and their reviewers.

Competent personnel who represent functions that are concerned with the particular design stage under review conduct design reviews at various (sometimes predetermined) stages of the design and development process. The key element in design

---

TABLE 2.13 Design Review

---

21 CFR	Key Questions	Your Company's Response (Yes/No/Maybe)
820.30(e)	Are design reviews planned for all designs undertaken within the organization?	
	Do procedures ensure that these design reviews are conducted and the results recorded?	
	Was there a conscious effort to eliminate any potential conflict of interest situations in selecting reviewers?	
	Are all functions concerned with the design and development stage under review represented at the design review?	
	Are records of design reviews taken and maintained?	

---

review is the independence of reviewers and design and development team members. This, in fact, is the principle behind quality audits and assessment. Independent eyes and ears are not "biased." Quality system procedures must ensure that these reviews are formal, documented, and maintained for future review. Key questions concerning design review are listed in Table 2.13.

Design Review Considerations

Formal documented reviews of the design results are planned and conducted at appropriate phases of the design and development work. Such phases are defined by the design and development plan or the design change plan.[14] Reviewers should have no direct responsibility (maintaining their independence). The key characteristics of such reviews are the following:

- Documented (formal)

- Comprehensive (technical)

- Systematic examination (planned, logical steps)

- Evaluate capability of the design and identify problems (do not sympathize with the development team)

Practical Needs and Added Value of Design Review

The value added by a design review comes from having an independent body of peers reviewing the design (with a "different set of eyes"). This is especially valuable

---

14. In the latter case, design reviews apply to a product already on the market that is being exposed to a design change.

when the review team has a strong foundation in customer wants and needs, reliability, safety, product and process technology, company quality systems, and industry regulations. Note that the design and development of products and processes is an iterative work. Therefore, identifying problems, issues, and opportunities is an expectation of the review process. During design reviews, an assessment of the progress (or lack of it) must be conducted so that the design and development team can move to the next phase of the project or go back and address the issues identified.

Elements of an Effective Design Review

In an effective design review, the problems and issues are:

1. brought forward,

2. discussed in detail,

3. analyzed,

4. assigned to a team member (corrective action),

5. followed up (corrective action effectiveness), and

6. closed.

In addition, the final outcome of each review is also documented as appropriate (in the DHF). Chapter 4 provides examples and guidelines for the development of procedures for design reviews.

Practical Implementation of Design Review

The implementation of a design review should involve a design review coordinator, leader, or chairperson who is independent from the design and development activities. The initial responsibilities of this individual are to enroll qualified technical design reviewers and to define the rules to be followed. Having a design and development team roster with their reviewers is a sound and common practice in industry.

The deliverables of the design and development team members responsible for specific activities are assessed by "expert reviewers" on a regular basis, forming a continuous flow of information and feedback loop. This is because it makes no sense to wait for the formal design review if issues can be resolved at a lower level and to a deeper extent. These "desk reviews" by peers are usually called verification activities because they are not typically comprehensive, definitive, or multidisciplinary in their scope. However, significant issues may be identified for the formal design review (multidisciplinary). This is a crucial practical principle.

We have seen some medical device companies go through a lot of pain when first implementing design reviews. One reason for this distress is the indifference of some reviewers to a project's progress, especially reviewers from functional areas who are not familiar with the specific design and development work.[15] Reviewers should be committed, technically competent, and well acquainted with the design under evaluation. They should also be familiar with the design and development plan or design change plan and the project's jargon. Each individual design and development team member should bear self-responsibility for keeping his or her reviewer(s) up-to-date regarding the project's progress, resolving issues, and ensuring specific deliverables, including documentation.

After the first design review, the coordinator, leader, or chairperson ensures the closure of corrective actions or resolution to conflicts and issues. In fact, this must be the key information to start the next design review. Ideally, prior to the next review, all reviewers and design and development team members should be notified of the corrective actions and/or issue resolutions because such corrections and/or resolutions may possibly create new ones.

How Many Design Reviews Are Required?

The regulation does not define any number of required design reviews. It does, however, make the following implications:

- As many as needed

- Defined on the design and development plan or in the design change plan

- Will vary from product to product, according to complexity

Myths About Design Reviews

The following myths about design review are illustrative of misconceptions of this important function.

- *Upper management must be part of it.* If upper management has the technical knowledge and competence to evaluate and challenge a design, then it can participate. Otherwise, no value is added to the review process by including upper management. Reviewers are supposed to be technical experts capable of understanding and challenging the science and engineering of the medical device under review.

- *Only people from R&D or New-Product Development can be reviewers.* Depending on the stage of the project, personnel from different departments (sometimes even

---

15. A way to handle this management dilemma is to include accountability and recognition for design reviews as part of the individual's performance review process.

outside experts) must participate. Again, the key question to be answered is whether their review adds value to the design and development process.[16]

■ *Design reviews are not needed for changes to existing products.* If the changes alter the design of a device, then the quality assurance group should evaluate the possibility that such changes may alter the indications, contraindications, and intended use, and/or may add new hazards. Therefore, the change can be seen as a "redesign" that requires not only a design change plan that includes design reviews, but many other elements of design control as well.[17]

■ *The entire team of reviewers must participate on every design review.* In complex projects, such as the design of capital equipment that includes embedded software, the creators of the design and development plan may decide that there will be specific design reviews just to evaluate software. In these reviews, those involved in the development of the software as well as systems integrators and their reviewers might be the only individuals needed for an effective design review.

Design Verification

The objective of the design verification requirement is to confirm by examination and show objective evidence that design outputs meet design inputs. The company should ensure that competent personnel participate in this activity. These activities must be planned and routinely examined, and the results must be documented for each design stage. Table 2.14 lists key questions for design verification. Common sense has driven some companies to adopt similar concepts, such as "engineering pilot," "design pilot," "engineering build," "qualification runs," and so on. The regulation aims at providing a sense of formality (i.e., procedures) and structure (i.e., design plan[18]) within the DHF.

Relationships Among Design Input, Design Output, and Design Verification

What can we learn from the example of design input, output, and verification shown in Table 2.15?

---

16. This represents a key challenge to quality managers and quality specialists. Only documentation requirements are met if the quality professional cannot understand the technology involved. The quality personnel involved in design reviews ideally also have technical depth.

17. Design change plans are discussed in Chapter 4.

18. Beyond the generality of the design plan, performance, quality, and reliability goals should already be established. Some firms may decide to include all the project requirements in the design and development plan; others may decide to establish interdependent quality and reliability plans in addition to the design and development plan. The same would apply to the design change plan.

TABLE 2.14 Design Verification

21 CFR	Key Questions	Your Company's Response (Yes/No/Maybe)
820.30(f)	Do you plan, establish, document, and assign design verification to competent personnel?	
	Is design output compared to design input to ensure that requirements are met at each design stage?	
	Do you undertake qualification tests and demonstrations?	
	Are alternative calculations carried out?[a]	
	Are new designs compared with similar proven designs when applicable?	
	Do procedures ensure that design stage documents are reviewed and authorized before release?	

[a]For example, some technical groups develop analytical models for a design and then computer simulations as well as prototype simulations are executed. In reality, three models are trying to predict the same outcome.

- The design output column is really a set of manufacturing process steps (DMR).

- The design verification activity will be part of the DHF. This documentation will provide the DMR with scientific rationale or backup.

- The design verification activity is in reality part of a reliability plan (i.e., the connection to other company quality systems).

- The safety factor of 3 is imminent information for the risk analysis.

In typical day-to-day operations in the medical device industry, cost reduction ideas must be generated. Usually, a change in materials and processing are typical sug-

TABLE 2.15 Example of Design Input, Output, and Verification

Design Input	Design Output		Design Verification
	Design Specification	DMR	
The medical device will be used in trauma rooms. It must be capable of withstanding adverse conditions (e.g., accidental pulling by the tubing).	The bond strength between a luer lock and tubing (IV line) should withstand p pounds of axial force without detaching from the tubing.	The raw material for the luer lock will be X and the solvent Y. Before inserting the tubing into the luer lock, the solvent will be applied, and a curing of T minutes will be allowed.	At 99% reliability and 95% confidence, a safety factor of 3 was obtained during a stress–strength test.

gestions. The authors hope the example presented here brings awareness about the need for the engineers and scientists in operations or manufacturing to know the rationale behind the "specs." Operations must have full access to the DHF and competent personnel who can take a cost reduction idea, or process deviation, can analyze its impact in design inputs, and be able to correctly verify and validate a design change.

Typical Design Verification Activities and Overlap with Design Output

Design verification is performed throughout the entire process of design and development. It mainly involves testing, simulation, inspections, modeling, and analysis. A key element of effective design verification has to do with test and measurement methods and manufacturing capability. As part of the design and development activities, the team is supposed to define appropriate design characteristics that actually describe its performance as well as its degree of conformance to design inputs. This is the time when test method requirements must be identified and validated. Such attributes as sensitivity, resolution, linearity, and so forth are used to specify test methods. In the case of dimensions, GR&R should be performed to ensure repeatability and reproducibility. The design and development team is supposed to identify what percent of the total specification range can be "consumed" by the error of measurement (see Figures 2.3 and 2.4).

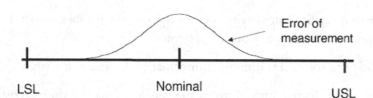

Error of Measurement as Percentage of Total Specification

FIGURE 2.3 Poor error to specification ratio.

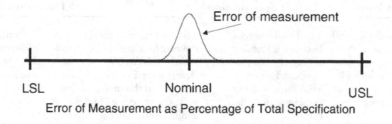

Error of Measurement as Percentage of Total Specification

FIGURE 2.4 Good error to specification ratio.

The test method validation and GR&R should be executed once FMEAs are available because the allowable percentage of error of measurement should be related to the risks involved in accepting a rejectable dimension (e.g., a false positive). Most of these design verification activities become DHF records that sustain the effectiveness of design outputs, such as the following:

- Risk analysis and management results

- FMEAs and FTA

- Test method requirement specifications

- Test method validations

- GR&R

- Reliability prediction results

- Component derating

- Design equivalency analysis

- Biocompatibility and toxicity results

- Packaging feasibility and results

- Software verification

- Transit test

- Third-party certifications (e.g., IEC/AAMI standards)

Design Verification Outcomes

Table 2.16 depicts four possible general outcomes of design verification. It shows that a failure in verifying the design outputs against design inputs when the input is wrong may lead the design and development team into additional iterations—perhaps into redesign—without knowing that they should fail design validation. If the verification passed, but the input was wrong, there could be false celebrations because the design validation should fail.

If the input was right, a verification failure means additional iterations until output equals input. If output equals input when inputs are right, then design validation should pass.

Design Validation

The objective of the design validation requirement is to confirm by examination and show objective evidence that the final design output consistently meets the specific intended use. Design validation always follows successful design verification. Because

TABLE 2.16   Potential Outcomes from Design Verification

		Design Output	
		Wrong	**Right**
Design Input	**Wrong**	Team will know only that the design does not work. If redesign takes place, it should fail in design validation.	Lucky you! At least for now.
	Right	Redesign.	Good job, now let's see the design validation.

design verification is done while the design work is being performed, the medical device may not be complete or may not be in its final configuration. For design validation, however, the team needs to have the final medical device. If the product has more than one use, multiple design validations may be necessary. Key questions regarding design validation are given in Table 2.17. Figure 2.5 depicts the relationship between design validation and design verification.

Design validation includes software, hardware, and a hardware-software interface by challenging the source code in its actual use conditions. For example, whereas embedded software verification is done by emulation of the source code, software validation is done once the software has been "burned" into the chip or electronically programmable read-only memory (EPROM) and the system is challenged.

Design Validation Strategies

One conservative but very complete strategy is to use medical devices already exposed to worst-case conditions for design validation. For example, a plastic device used in surgery is sterilized twice for simulation of "resterilization." It is then sub-

TABLE 2.17   Design Validation

21 CFR	Key Questions	Your Company's Response (Yes/No/Maybe)
820.30(g)	Is design validation used to ensure that the product conforms to the defined user needs?	
	Is design validation undertaken after successful design verification?	
	Is validation performed on products manufactured or assembled from initial production runs?	
	Is validation performed on the finished product under defined operating conditions?	
	Do procedures ensure that validation is carried out for each of the product's defined uses?	

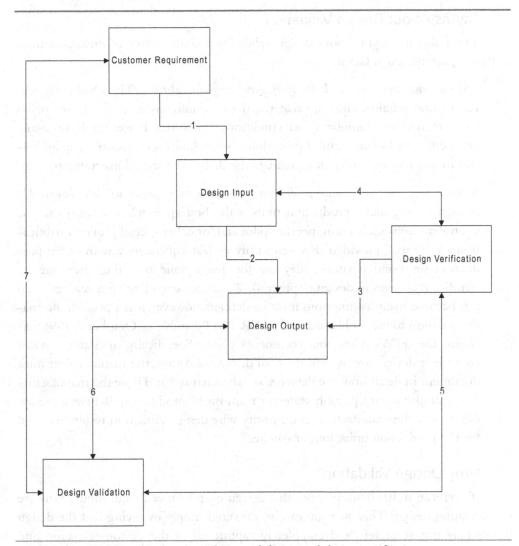

FIGURE 2.5 Relationship between design validation and design verification.

jected to shock and vibrations in its package as part of the transit test and packaging validation. It is then used for design validation. Successful validation then is meaningful because we have used a worst-case device that has been exposed to overstress and still does its intended function.

The regulation states that design validation should be performed on initial production units. Initial production units can be defined as units made from the PQ during process validation. Thus, the two design control elements need each other—the design transfer is not complete until successful design validation. More details are given in Chapter 3, which discusses the relationship between design transfer and process validation.

Myths About Design Validation

The following myths about design validation are illustrative of misconceptions of this important design factor.

- *The only way to validate a design is by performing clinical trials.* The Quality System Regulation requires ensuring that the device conforms to defined user needs and intended uses. Simulating use conditions can do this. For example, for many endoscopic and open surgical procedures, a preclinical evaluation using an animal model may be sufficient to validate the design of surgical instruments.

- *If the team cannot use initial production units, then they cannot validate design.* Although having initial production units is the best approach, the team can use equivalent units such as engineering pilot and/or other special process qualification run units,[19] provided they can justify a clear equivalency with initial production units and whatever they use for design validation (i.e., there are no significant changes in design or process). We do not encourage our readers to do this because many assumptions must be defined; however, it is a possible alternative solution to the problem. The 21 CFR 820 Preamble of October 7, 1996, explains the FDA's view on commentary 81. Specifically, it states, "when equivalent devices are used in the final design validation, the manufacturer must document in detail how the device was manufactured and how the manufacturing is similar to and possibly different from initial production. Where there are differences, the manufacturer must justify why design validation results are valid for the production units, lots, or batches."

Why Design Validation?

If verification has demonstrated that design outputs met design inputs, why are we validating design? This question can be answered simply by saying that the design inputs may not accurately and completely capture all of the customer's wants and needs. Also, even if design inputs are right, design outputs could be wrong. One possible reason for this could be changes in customer requirements since the design was initiated. If design inputs and outputs are right, a problem could have developed when the design was transferred to manufacturing. This is the main reason why initial production units are the best choice for design validation. Notice that design validation is a final challenge to all the existing quality systems, including design control, training of manufacturing personnel, process validation, and so on. In fact, because design transfer requires process validation, the design control activities are not completed un-

---

19. Engineering pilot or qualification runs are typically done in a manufacturing setup with special controls and under scrutinized surveillance by the design and development team (e.g., process development). The potential limitations of the devices made here are that the manufacturing process may not be the final one as well as the fact that in real life the team will not be "babysitting" manufacturing operators.

til after successful process validation. Design validation is basically the main and final step in releasing the product to the market.

If, on the one hand, the output from the process does not meet customer needs and intended use, the manufacturing process becomes worthless. But if, on the other hand, the process is not repeatable or reproducible, the design validation is also worthless because no manufacturing process can ensure equivalent performance from unit to unit and/or from batch to batch.

Design Validation Outcomes

Table 2.18 depicts four possible general outcomes from design validation. Note that a failure in validating the design against customer wants and needs when those wants and needs are wrong may lead the design and development team into additional iterations—maybe endless redesign—without knowing that the problem lies with the assessment of the customer's wants and needs. It is very unlikely that if the user's wants and needs are wrong, the validation will pass.

If the user's wants and needs assessment is right, a validation failure means additional iterations because the design inputs and/or outputs may not have been correctly defined. If the design validation passes when the user's wants and needs are right, then you have a medical device.

Differences Between Design Validation and Process Validation

Many companies are still confused about the differences between design validation and process validation. Basically, design validation ensures that the device meets its predefined intended use. It is an answer to the question: "Have we designed a device that works per customer needs and expectations?" Process validation ensures that the manufacturing process is capable of consistently producing a device that meets its predetermined specifications.

Personnel in manufacturing operations know that when they make a change to a process, they must evaluate the possibility of having to validate such change. How-

TABLE 2.18 Potential Outcomes from Design Validation

		Medical Device	
		Fails Design Validation	**Passes Design Validation**
Customer Wants and Needs	**Wrong**	This can become an endless loop. The design and development team may continue redesigning without noticing incorrect design inputs.	Lucky you! You may win the lottery!
	Right	Redesign; the outputs might be wrong.	Good job, you have a product.

ever, in many instances no evaluations of the impact to design are done because "the process was validated." This is a misperception of both the regulation and common-sense engineering and science practices. The typical argument is that the change still allows the product or output to meet such specifications as dimensions. However, manufacturers must keep in mind that not all the relevant product specifications are inspected or tested per the DMR. Some characteristics of the process output may have been only verified and validated during design and development. For example, suppose that reprocessing or reworking certain plastics and metallic components in a device could affect reliability. The change may thus yield less physical strength. If the firm has not instituted reliability testing as part of the day-to-day process (e.g., DMR), this process change, or deviation, may not be correctly assessed.

Design Changes and Design Transfer

All design changes must be authorized by people responsible to ensure the quality of the product. Procedures should be established for identification, documentation, and review of all design changes. Table 2.19 lists key questions about design changes. Design controls and other quality assurance activities must follow the same rigorous procedure as that adopted for the original design. But, what kind of design changes are we talking about?

Four typical situations are involved in controlling design changes. The manifestation of design changes can come during the design and development phases or after the product is on the market. Also, the design change can merely be a documentation correction or it could mean a change in the physical configuration of the device. The matrix in Table 2.20 depicts the impact of design changes, presented in the following typical situations:

- *Situation 1:* Document control is a straightforward classical GMP quality system for existing products. It is aimed at enumeration, identification, status, and revision history of manufacturing specifications, as well as testing instructions, BOM, and so on (i.e., all elements of the DMR).

TABLE 2.19   Design Changes

---

21 CFR	Key Questions	Your Company's Response (Yes/No/Maybe)
820.30(i)	Do procedures ensure that all design changes and modifications are identified and documented?	
	Do procedures ensure that changes at the suppliers or contract manufacturers are also reviewed?	
	Do procedures ensure that all design changes are reviewed and approved by authorized personnel?	

TABLE 2.20   Design Changes and Product Life

	During Design and Development	After Product Has Been Released to the Market (Existing Products)
Document Control	Situation 2	Situation 1
Change Control	Situation 3—What is the impact on all other design outputs already approved?	Situation 4—Issue is design validation. How can a factory do this?

- *Situation 2:* Once the design controls are part of the firm's quality systems, the firm needs to control the documentation that is being "drafted" during design and development. Many temporary documents exist during the phases of product design; most of them will be subject to multiple changes. Thus, the big question concerns how the design and development team can ensure harmonization between the already approved design elements (e.g., design reviews) and new elements of change? Figure 2.6 depicts design changes with a diagonal line that implies multiple changes in this temporary or conditional DMR during the entire design and development life cycle of the device. It is important to realize that not only DMR, but also elements of design verification and validation, can be affected, including the DHF.

- *Situation 3:* Change control per se has to do with the physical characteristics of the device, its acceptance criteria, or its testing or evaluation methods. For products under development, a logical procedure has to be implemented to expose the entire design and development team as well as the group of design reviewers to the changes. The relevant questions here are: How do we ensure that the design outputs that have already been reviewed and approved are still valid or will not be affected by the change? How do we ensure harmonization with DMR elements already in place?

- *Situation 4:* A bigger challenge in terms of regulatory compliance and business risk is the control of design changes on existing products. The changes can alter not only the design, but the intended use of the product, and thus the 510K or PMA submission to the FDA. Another possibility is that the change might affect some other device or subsystem manufactured. Of greatest concern in this situation is the fact that manufacturing operations are typically the ones requesting the changes. Without competent personnel with access to and understanding of the DHF, how can approvers of change be able to make conscious decisions? Also, manufacturing operations may never have the means to execute a design "revalidation" upon design changes. This is where the firm should make use of

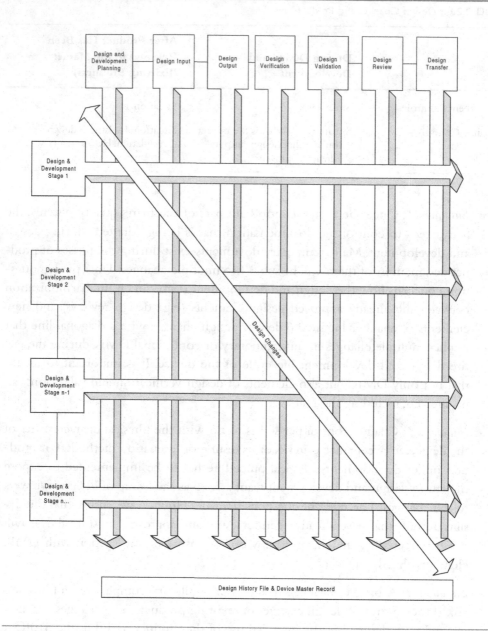

FIGURE 2.6 Design changes occur throughout all phases of design and development.

"design change plans," as described in Chapter 4. Like process validation, the application of design controls never ends. It applies to design changes to both the process and the product, even when the product has already been launched to the market. Controlling design changes is the biggest chunk of the role of the factory in design control. In our experience, this is still a very weak point in most companies. An alternative solution is to have permanent highly technical personnel on the factory premises with a "no limit" access line to R&D. Also, the factory should have full access to the DHF, especially the risk analysis and management report and the FMEAs.

Role of the Factory in Design Control

The most frequently asked question by management and technical personnel working in manufacturing or operations facilities is: "Are the FDA's design controls applicable to us?" Two main reasons exist for this paradigm. First, many companies believe that they have separated R&D from the manufacturing facilities. Thus, they still have to understand the scope of the concept "design and development." Second, many of those factories have existing products, and thus they believe the regulation is not applicable. Chapter 3 makes clear that design control does apply to manufacturing facilities. In this section, we emphasize the fact that many changes initiated by manufacturing operations, or the factory, are indeed design changes, and, as such, they are required to follow design control procedures.

The purpose of Table 2.21 is to provide a checklist that defines whether a given proposed change to the DMR may affect design. A "yes" answer to any of the questions posed in Table 2.21 is an explicit admission to the world of design control. The key element of control is to have a company procedure that evaluates potential impacts to the device's intended use and calls for a revalidation of design, an equivalency analysis, and/or product performance qualification when necessary.

Companies have to be careful about design changes because any of them could automatically trigger a 510K or Premarket Approval resubmission. The regulatory affairs and compliance team is always interested in analyzing and approving all design

TABLE 2.21   DMR Change Request Evaluation Criteria

Within Your Factory Operations, Are There:	Yes/No
Changes to the product (e.g., system configuration, packaging, performance-related features, and device functionality)?	
Changes to the process or process parameters?	
Changes to the specifications (e.g., in process, of materials)?	
Changes in materials?	
Changes in process conditions (e.g., humidity, light intensity)?	
Changes in components?	
Changes in sterilization methods?	
Changes in the product's software?	
Changes in the state of knowledge upon a field action?	
New products based on an existing platform?	

TABLE 2.22 Typical Factory Situations and Regulatory Concerns

Typical Situations in the Factory	Typical Concerns from the Regulatory Affairs Team
We will make changes or deviations to the process.	What are the short- and long-term effects on product safety and effectiveness?
We will expose the initial production units to an accelerated stability protocol.	Is it a validated test method? Is it applicable to this specific product? What are the risks involved? Is there any correlation with real-time stability?
We will show that the product still meets its intended use. Therefore, equivalency will be demonstrated.	How about life expectancy (reliability estimates)? What kind of overstress model was used during design and development? What were the overstress conditions? You can show equivalency now, but what about in the long term?
Platform Changes[a]	
Changing the materials is not significant. "We need to reduce costs."	Will the results from biocompatibility be equal? How about the stress-strength relationship?
For in-vitro diagnostic (IVD): The new analyzer can use the same reagents as the current one.	Will the assay have the same analytical sensitivity? What about functional sensitivity?

[a]A platform change occurs when a new product is made based on an existing one. Although the result is a new product design, a very significant amount of the design elements are the same. For example, a new bipolar electrosurgical forceps can have a design identical to an existing one, but the tip of the poles has a 20 percent wider area of contact. Are there any new hazards now?

changes. Table 2.22 presents a list of typical "factory situations" in which the linkage to design control requirements is imminent.

Factory personnel must bear in mind that even slight process changes can lead to design changes. An indicator of whether a change can affect a design can be achieved by merely focusing on the intended use, design inputs, FMEA, and risk analysis. Thus, once again and throughout this book, we emphasize that *factory personnel should have access to the Design History File and should be competent enough to understand and interpret such records.* A typical strategy is to establish "technical services" teams in manufacturing with enough technical depth so they can think through and analyze situations like R&D personnel do. In large companies, these teams become permanent liaisons to R&D and corporate regulatory affairs in addition to planning and executing many of the activities pertaining to design transfer, such as process development, process characterization, process specifications, inspection specifications, process validations, and so forth (for more details, see Chapter 3). A typical paradigm of concern is the wrong belief by corporate groups that the DHF should not be shared with manufacturing operations because it contains proprietary information such as patents. But how can a manufacturing technical services group be able to evaluate a deviation without knowing the design considerations of the device? Conversely, design and development activities include the manufacturing process!

Design Transfer

For in-depth information on design transfer, see Chapter 3.

Design History File

The DHF is the compilation of evidence that shows how the design was developed in accordance with design controls, specifically the design and development plan or the design-change plan. As the name indicates, the DHF compiles the history of the design. This file should be made available to manufacturing operations. In essence, analysis and review of this file should be the first step taken by the process validation specialists. Chapter 4 provides some guidelines for building the DHF.

Myths About the DHF

The following myths about the DHF are illustrative of misconceptions of this important design control element.

- *The DHF has to be a big, thick binder.* The DHF is really an umbrella concept. In a small firm, it can be physically assembled in a binder. In bigger firms with very well developed quality systems, it may be just a concept. For example, design verification, design validation, process validation, reliability testing, and so forth can be well-documented protocols stored in the "GMP records room." However, a "book" or some equivalent keeps records of the identification numbers for all of these studies. The same applies to risk analysis, FMEAs, and other elements of the DHF. A design history matrix is very helpful (see Table 4.4 in Chapter 4 for an example).

- *Manufacturing does not need a DHF.* This is a big mistake, and, technically, it forbids manufacturing from initiating design changes unless there is a permanent contact in "R&D" available to do so. We have seen this happen, especially with firms whose manufacturing operations are in distant locations (e.g., tax havens or emerging economies). The usual excuse is the DHF contains proprietary information. The question is, then, how will they justify the design change?

Further Reading

ANSI/ASQC. 1995. *D1160-1995: Formal Design Review*. E-Standard. Milwaukee: ASQ. (May be purchased on-line at http://e-standards.asq.org/perl/catalog.cgi?item=T218E)

GHTF. 1999. *SG3-N99-9: Design Control Guidance for Medical Device Manufacturers*. June 29. Global Harmonization Task Force. (May be downloaded at http://www.ghtf.org/sg3/inventorysg3/sg3-n99-9.pdf)

Kuwahara, Steven S. 1998. *Quality Systems and GMP Regulations for Device Manufacturers*. Milwaukee: ASQ Quality Press.

CHAPTER
THREE

Design Transfer and Process Validation

This chapter describes the relationship between design transfer and process validation. It offers a practical definition for design transfer and describes the role of the factory in design transfer as part of design control requirements. It then presents a practical "how to" road map of process validation and a practical definition of this concept. Both process validation and design control are seen as "never-ending" activities.

Successful design transfer is more than releasing the DMR to the manufacturing division. As depicted in Figure 3.1, design outputs (the design) are translated into manufacturing specifications (the DMR). The project enters then into the process validation cycle that will analyze, challenge, complement, and improve the DMR. As part of this cycle, the design will be validated, and if design outputs, including the design of the manufacturing process, meet customer needs, then design transfer closes its loop.

A general flowchart of process validation and its interdependence with a company's quality systems is presented later in this chapter. The chapter also presents the Global Harmonization Task Force (GHTF) process validation guidance and presents some strategies for the whole validation program. Concepts such as installation qualification (IQ), operational qualification (OQ), and performance qualification (PQ) are discussed as well.

Guidelines for validation of processes associated with a new product[1] are presented and discussed in detail because validation is one of the most misused concepts in the medical device industry. This chapter attempts to clarify the concept and to highlight many fallacies about it. Throughout this chapter, it will become obvious that good procedures and good execution are beneficial not only because they help the firm comply with the regulation, but also because they increase the firm's profit margin.

---

1. Although not less important, facility, utilities, and sterilization validations are not discussed here because they are not necessarily product specific.

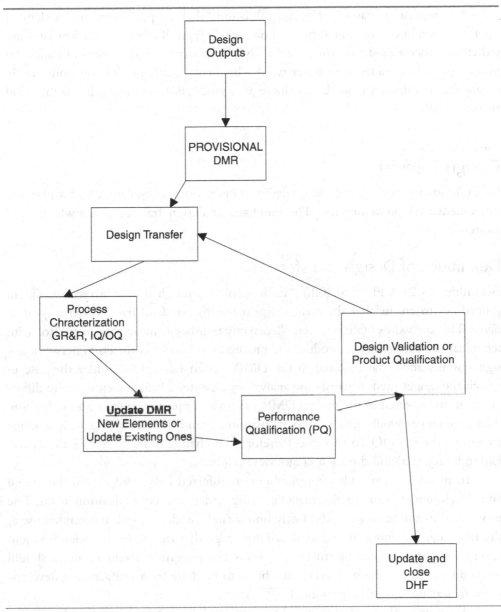

FIGURE 3.1 Design transfer flowchart.

Relationship Between Design Transfer and Process Validation

In essence, the concepts of design transfer and process validation have many common-alities. The design and development cycle can be said to end with appropriate process and design validation. In this context, process validation is an activity within design transfer.

The regulation states that design validation "shall be performed under defined operating conditions on initial production units." Typically, first production runs are either the process validation runs, specifically PQ, or manufacturing runs just after the process has been validated. In other words, the term "initial production units" really means the first devices made under the regular production system for human use and market supply.

Design Transfer

In the following section, we define design transfer from a regulatory and a practical implementation point of view. The emphasis in design transfer is knowledge and awareness.

Definitions of Design Transfer

According to 21 CFR 820.30(h), "Each manufacturer shall establish and maintain procedures to ensure that the *device design is correctly translated into production specifications.*" The key words (italicized here) "correctly translated" mean that an auditor with technical knowledge of the product and process could find a connection between design outputs and what is stated in the DMR. As an example, consider the case in which the ranges used in worst-case analysis by an external manufacturer were different from those stated in the drafted DMR due to an opportunity for a cost reduction. The new ranges would imply an extrapolation beyond the worst-case analysis range examined for the OQ. In this case, therefore, the change order to release the DMR had to be rejected and the right ranges were put back.[2]

In practical terms, what is really being transferred is *knowledge* from the design and development team to the manufacturing and/or process validation team. The process validation team must absolutely understand the device and its intended use as the first step. The amount and kind of knowledge that the design and development team have about the manufacturing process is also relevant. Careful attention should be paid to what is done by R&D and what is to be done by manufacturing development or process validation personnel.

As an example, we can think of process development in two stages. First, consider the design of a new manufacturing machine or piece of equipment. The design of the new machine is typically done by a design and development team, sometimes in combination with a consultant or with the machine manufacturer. But designing a manufacturing machine is not synonymous with process characterization.[3] That is, a lot of unknown behavior is associated with a newly designed machine. Through de-

---

2. This was a "GMP certified" contract manufacturer.

3. Process characterization is the comprehensive understanding of how people, materials, equipment, procedures, methods, and environment impact the manufacturing process.

sign of experiments (DOE)[4] and other statistical tools used during process characterization, technical manufacturing personnel can really learn what input parameters affect the output characteristics and in which way. In other words, if the manufacturing personnel know what the input parameters[5] (independent variables) are and how they affect the output characteristics[6] (dependent variables), they may have a way to control the manufacturing process.

Design Transfer Questions and Related Quality Systems

The key question related to design transfer is shown in Table 3.1. Other quality systems related to design transfer include the following:

- Process validation

- IQ/OQ/PQ

- Test method validation

- GR&R

- Manufacturing systems software development and validation

- Statistical techniques

- Process control methods

- DMR preparation and review

- Document (DMR) review with document users

Design Transfer Summary

In summary, when new manufacturing equipment is designed, two main development steps are involved: the first is the design, per se, of the equipment, and the second is characterizing the equipment or machine. Typically, the second step is a function of the "pilot plant."[7] If a company is operating in a "direct to site" mode,[8] then the sec-

---

4. Manufacturing processes and equipment do not behave according to deterministic laws of physics. Their prediction models may be based on physical or chemical principles, but in reality, they are empirical.

5. Input parameters really mean the machine or process settings that produce a product or output per predefined specifications.

6. Output characteristics are the product specifications that the manufacturing process and its settings are supposed to produce. They are also known as dependent variables because they depend on the process and its settings.

7. The idea of a pilot plant comes from the technical necessity of having manufacturing close to R&D, at least while awaiting "steady-state" production, which is characterized by achieving a state of flatness in a learning curve.

8. Thus, eliminating the concept of a pilot plant.

TABLE 3.1 Design Transfer

21 CFR	Key Questions	Your Company's Response (Yes/No/Maybe)
820.30(h)	Does your company have procedures to ensure that the device design is correctly translated into production specifications?	

ond development step will have to take place at the transfer site. This second development step falls under the definition of OQ, specifically under the concept of process characterization.

Design transfer may occur via documentation, training, R&D personnel sent to manufacturing, and/or manufacturing personnel having been part of the design and development team. All the design transfer activities should be listed in the design and development plan. However, training and documentation do not fulfill the whole purpose of design transfer. The expected results of effective design transfer are:

- The product has manufacturability and testability.

- The process is repeatable (item to item within a batch or lot, see Figures 3.2 and 3.3).

- The process is reproducible (lot to lot, see Figures 3.2 and 3.3).

- The process is under statistical control (stable), and thus it is *predictable* (such as seen in Figure 3.2).

- Manufacturing personnel *know what they are doing* and what process parameters need to be adjusted, as well as how, when, and why to adjust them.

- The DMR documents are adequate.[9]

- The manufacturing and acceptance specifications are realistic and meaningful.

- Raw materials and components perform as expected.

- Suppliers know what they are doing.

- There are no surprises.

In sum, *effective design transfer results in a manufacturing process that consistently ensures a medical device that is safe and effective.*

---

9. As opposed, for example, to manufacturing instructions that can be understood only by the development engineer who wrote them.

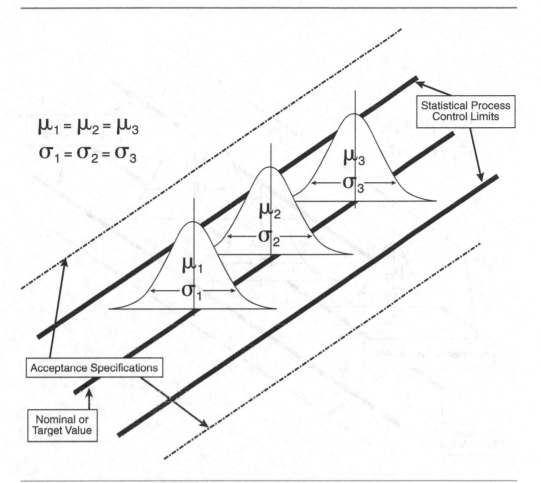

$$\mu_1 = \mu_2 = \mu_3$$
$$\sigma_1 = \sigma_2 = \sigma_3$$

Statistical Process Control Limits

μ_3

σ_3

μ_2

σ_2

μ_1

σ_1

Acceptance Specifications

Nominal or Target Value

FIGURE 3.2 Case in which process is stable and all three lots fall within acceptance limits. Process reproducibility is assessed by comparing the means. Process repeatability is assessed by comparing the standard deviations.

Process Validation

Many different kinds of medical devices exist, and they encompass different materials, technologies, and engineering principles. However, they all go through a manufacturing process that involves variables such as people, materials, methods, measurements, and equipment, among others. Process validation is the set of activities (e.g., analysis, experimentation, testing, evaluation, and confirmation) aimed at controlling the variables that affect the process. Real-life experience, however, shows that complete control is utopia. Otherwise, why would medical device companies typically have engineers and scientists "troubleshooting" and/or "revalidating" processes?

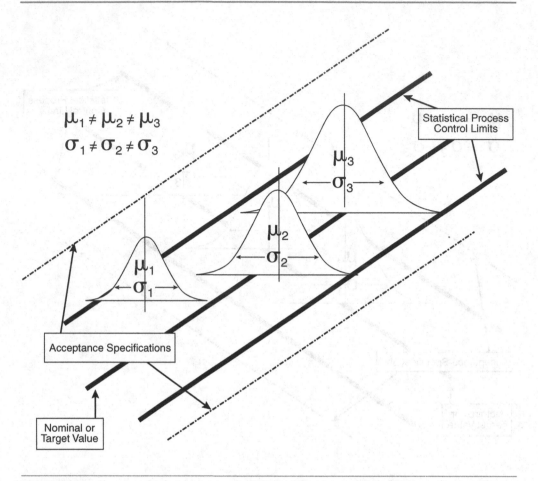

FIGURE 3.3 Case in which process is unstable, but all three lots fall within acceptance limits. Process reproducibility is assessed by comparing the means. Process repeatability is assessed by comparing the standard deviations (production run or lot number 1 has a better repeatability than number 3).

The benefit of effective design transfer and process validation is appreciated most when a hit to productivity occurs due to a process stop or process shutdown, or when batches are rejected, or—in the worst case—when a product is recalled. Thus, effective design transfer and process validation should be seen not as regulations, but as normal parts of the business strategy. A process can be validated by three approaches—prospective, retrospective, or concurrent:

- Prospective validation is the simplest approach of all. This is done after the product is designed and the process is developed. The new product is not sold before completing this validation.

- Retrospective validation is the most difficult method. It can be effective only if the process and the product have been monitored for all relevant quality characteristics. But how can we demonstrate that all relevant quality characteristics have been monitored? How can we rely on past data if we have process failures, field complaints, or heavy reliance on testing?

- Concurrent validation is the middle point between prospective and retrospective validation. It is the best approach for existing products and processes undergoing some kind of design change.

Qualification activities can use retrospective data while the whole validation cycle (see Figure 3.4) is being executed. Some of the experiments and decisions during process characterization and OQ could be based on such historical data. Validation runs are still required. The three kinds of qualification activities are IQ, OQ, and PQ. They are part of the entire concept of process validation. Thus, they are part of the process validation flowchart in Figure 3.4.

The Process Validation Flowchart

Process validation is, at the least, everything contained in the flowchart in Figure 3.4. Figure 3.4 has two main elements: process validation and the firm's quality systems. Process validation is an integral part of the firm's quality systems. In fact, all the quality systems are integrated to each other. The design and development team as well as the process validation team must know the firm's quality systems as part of their day-to-day job skills. Adequate planning is not feasible without the validation team members having this knowledge. During the PQ runs, especially if the output is salable, many of the firm's quality systems are being invoked and challenged. If PQ runs fail, then the corrective and preventive action quality system comes into play, as well as material control and segregation, for example. Typical reasons for failure are lack of training, incomplete bill of materials (BOM), conflictive specifications, confusing procedures, lack of process characterization, and lack of knowledge. Each of these potential reasons can be addressed by one or more of the quality systems a firm is supposed to have. This is, in fact, the message in Figure 3.4. The column to the left represents typical quality systems. From the beginning of this flowchart, the firm's quality systems are being invoked.

Learning Stage

The process validation team should start by getting acquainted with the device, the insert, the indications, the science behind the intended use, and the engineering, or technology, of the design. This is, in essence, the DHF. Companies that have understood the complexities involved in bringing innovative new medical devices to the market at a fast pace and in compliance with the regulation have implemented advance quality systems and have their own advanced quality sciences groups. In such companies, the R&D or design and development team includes process development and quality/reliability engineers who work with manufacturing engineers and/or validation specialists in the development, characterization, and eventual validation of the manufacturing process as well as in the design validation. Both process development and quality/reliability engineers provide that link between manufacturing and the DHF in a useful and practical manner. Learning the DHF by the manufacturing personnel is faster and can occur concurrent to many of the other activities going on.

Validation Plan   Once the learning step has taken place, a main outcome of this stage is a *validation plan*.[10] One of the most common root causes for troubles in process validation and design transfer is lack of planning. This plan typically considers the firm's "generic validations"[11] such as new facilities, utilities, and sterilizers (e.g., autoclave), as well as the new[12] product-specific activities (see Figure 3.4). This plan should be detailed and technically comprehensive. The typical mistake made by some organizations is to prepare a project management Gantt chart and ignore the technical challenges associated with process reproducibility and repeatability.

Process Map   The process map is a universal tool for communication and understanding what is involved. Medical device companies have two flows. The first is the manufacturing flow of materials and/or components used to build the device. This process map really shows what manufacturing steps are involved and in which order. The second process map is the flow of information that eventually becomes the DHR. The validation team should know both.

Equipment-Related Studies

During the equipment-related studies stage, the validation team "plays" with the equipment to learn how it works and what it does. Among the typical deliverables from this phase are the calibration and PM requirements, GR&R, and usually the IQ, among others. (See the separate section on IQ later in this chapter.) Here is where some engineering studies are performed and documented for the first time in the process validation project. Short runs or exploratory runs of the equipment are done to have a "flavor" of what is going on and to verify whether the equipment responds to control of the parameters.

---

10. Caution is advised with this term. The GHTF guideline mentions a "master validation plan." For some firms, the master validation plan is the overall plant-wide plan that includes all generic validations as well as all new product transfers.

11. The term "generic validations" is used in this book as a term that is not related to a new design or product, but to a shared entity, such as facilities and utilities that are shared by different products and/or a family of products.

12. Also applies to existing products undergoing a design change.

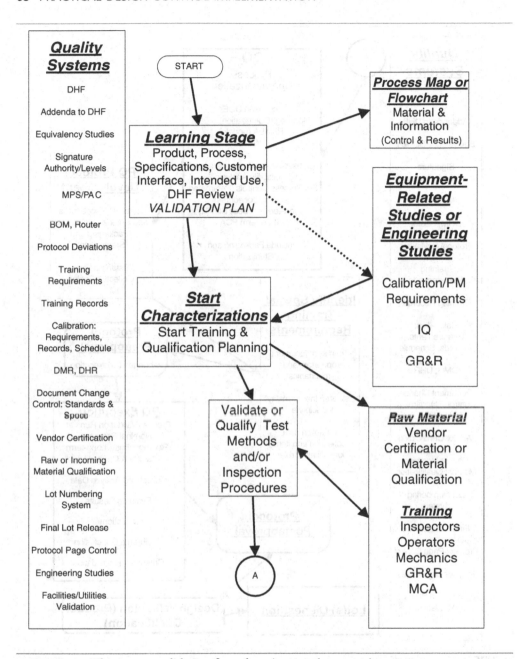

FIGURE 3.4 The process validation flow chart (*continued next page*).

Legend:

DHF = Design History File
DHR = Device History Record
DMR = Device Master Record
DOE = design of experiment
GR&R = Gage repeatability and reproducibility
IQ = Installation Qualification
MCA = measurement capability analysis
MPS = master production schedule

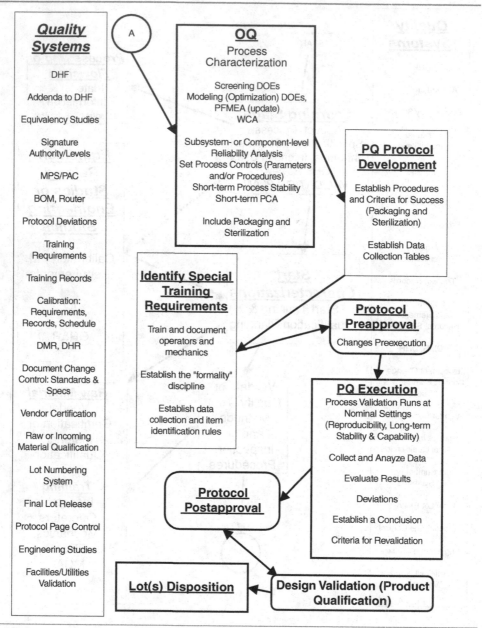

FIGURE 3.4   *continued*

Legend (*continued*):

OQ = operation qualification
PAC = Production Action Control
PCA = process capability analysis
PFMEA = process failure modes and effects analysis
PM = preventable maintenance
PQ = Performance Qualification
WCA = worst–case analysis

Start Characterizations

More understanding of the manufacturing process is needed in order to start developing OQ, other studies, and protocols. In the start characterizations phase, the validation plan should call for identification of training needed by the validation or design transfer team (e.g., new technology, new materials, new testing, new processes), the operations crew, and the QC/QA personnel, among others. Supplier evaluation is also planned and executed here. All this can be done in parallel with the engineering studies. Other quality systems, such as training and purchasing controls, are invoked. Assessing measurement requirements and capabilities is among the most important activities at this stage.

A typical question to be answered in this phase is: Does the firm have the human and physical resources to measure what the current DMR says must be measured? Measurement systems have to be addressed before OQ and other evaluations of the process because if DOEs and capability studies are done using imprecise or inaccurate measurement techniques, the process characterizations, capability studies, and related measurements will be worthless. Also, consider the issue of how good your product would be if the inspection and/or testing methods do not detect a nonconforming product.

Material Qualification

Large companies have a master list of approved materials from certified suppliers. In any size firm, those responsible for process validation must ensure that the right raw materials are used in the OQ/PQ activities and that the same raw material used in validation is also used for market production.

Test Methods Validation

The results from process characterization and eventual PQ are as good as the methods used to evaluate the process and the product. A significant error of measurement may imply a high number of false positives and/or false negatives when evaluating product and also it may lead to the wrong settings of process parameters. Test methods, measurement procedures (e.g., metrology), and inspection methods are typically not well understood in the medical device industry. In our experience, from 25 to 50 percent of design transfer problems are related to test methods and/or metrology issues such as error of measurement. Figures 2.3 and 2.4 in Chapter 2 illustrate this concept.

The DHF should indicate the product characteristics to be measured. It should also include a control plan that states the qualifications needed in a test method (e.g., precision, accuracy, sensitivity, linearity, resolution). A test method validation procedure is necessary in most firms. Test method validations must be done prior to starting process characterization (OQ).

Inspection Procedure Validation   If the inspection is done by automatic vision systems, this procedure requires an equipment and software validation.[13] A very good way to validate the entire system is by simulating good and bad parts and using statistical and probability models to justify the validation. That is, like a manufacturing process or a measurement device, the manufacturer should demonstrate that an inspection method is repeatable, is reproducible, and that it has the required discrimination capabilities.

Installation Qualification

The definition of IQ per the GHTF is "establishing by objective evidence that all key aspects of the process equipment and ancillary system installation adhere to the manufacturer's approved specification and that the recommendations of the supplier of the equipment are suitably considered."

The equipment supplier's recommendations are to be considered, but the responsibility for correct installation belongs to the medical device manufacturer.[14] Who can guarantee that the equipment's manual is correct?

The practical definition of IQ is "to ensure correct installation of any equipment used in any of the processes of the firm." Among other functions, the equipment can be used in:

- production,

- inspection,

- packaging,

- labeling,

- cleaning,

- testing, or

- sterilizing.

The term "equipment" includes, as examples:

- forming equipment such as stamping machines;

- injection molding machines and their molds;

- cleaning equipment, such as a tumbling machine;

---

13. Typically, inspection procedure validation is a weak point for many medical device companies.

14. This is why the medical device manufacturer needs to establish purchasing controls and procedures to qualify equipment vendors.

- processing equipment, such as a sonic welder; and

- custom-designed equipment.

  IQ seeks to answer questions such as the following:

- What is the availability of equipment documentation such as drawing(s), manuals,[15] and other controlled documentation?

- Are there instructions for use, maintenance (e.g., PM), and calibration?

  □ Is the equipment cleanable?

  □ Does anybody know how to clean the equipment?

  □ Are any instructions available?

  □ What about the possibility of carryover residuals—can they damage future production?

- How can the equipment functionality be verified against specifications?

- How can this equipment do automatic inspection (e.g,, programmable logic controller [PLC], visual or mechanical fixtures)?

- Is there any associated software (e.g., firmware)?

  □ How will this software be validated?

  □ How is it controlled?

  □ Which revision level do we have? Does it have a manual? How about a troubleshooting manual?

  □ What do we do if the vendor changes the revision level for the software? How are we going to know?

  □ Is it custom made? Who generated the software requirement specifications?

  □ Is any special hardware ancillary equipment or software needed (e.g., storage, databases)?

- Is the equipment stable?

- Are the lighting and flow of materials adequate?

- Is the product to be made light sensitive?[16]

- Can the equipment lubricants (e.g., oil, grease) contaminate the product to be made?

---

15. A typical mistake by some firms is not to safely store the equipment's manual. Another is not to have a procedure to respond to updates or upgrades (e.g. manuals, software modules, errata sheets) by the vendor.

16. Such as some tracers used for IVD.

- How do we ensure correct calibration?
  - Does the firm have the capabilities to calibrate the equipment?
  - Do we need a custom-made gage?
  - Are any personnel trained in such calibration?
  - If we need an outside calibration firm, how do we know they are qualified?
  - How about equipment controls?
  - How is the frequency of calibrations determined?
  - Has the equipment been entered in the calibration and PM databases or systems?

- What about tooling (e.g., training of operators, maintenance and repair technicians)?

- Can the equipment be repaired in-house? What would the required qualifications of an outside service supplier be? Are we covering this supplier with purchasing controls?

- How are we going to handle equipment spare parts and components?
  - What are the logistics?
  - Should the firm have spare parts in house?

- Are there limitations, such as products to be made or kind of raw material it can process?

- Are there potential issues with safety, ergonomics, or human factors?

Correct installation includes evidence that the equipment can safely work. Examples of an acceptable environment are: a clean room or just a controlled room, no issues with the plant layout,[17] no potential issues affecting equipment performance,[18] and proper setup.

---

17. As an example, after long weeks of planning and evaluation of a process consisting of three work stations, a contract manufacturer's IQ was rejected by the plant engineer because the third work station would block the air conditioner's return opening on the wall. This caused a delay of three weeks.

18. For example, an analytical balance should always be placed on a very rigid and static surface. In one case, an analytical balance was used to weigh plastic injection molded parts. Once in a while, there was an "out of control" point in the SPC chart. The IQ had ignored the fact that the analytical balance was directly under the exit grill of the heating, ventilation, and air-conditioning (HVAC) system.

Experts   As part of the IQ activities, firms should include subject matter experts. For example, consider the following:

- Have the safety officer examine the equipment and look for safety, ergonomics, and/or human–factor–related hazards.

  - For example, the equipment may be fine, but the metallic parts coming out may have very sharp edges or flashes that could wound the manufacturing operators.

  - Some examples of hazards are moving components; wiring exposure; incorrect grounding; inappropriate ventilation (e.g., heat sinks and/or fans); sharp edges; lack of emergency stops; potential for "projectiles"; incorrect connection to utilities, wiring, air, power, water, or other support systems.

- Determine whether this equipment uses shared[19] utilities, and what the impact would be to the operation.

Other IQ Considerations

- Is there any effect when increasing batch sizes?

- Can the equipment pass through existing doorways?[20]

Myths and Realities About IQ   The following myths about IQ are explored in this section.

- *We use a checklist.* Many companies work with a "pre-built, boiler plate, or template" protocol for IQ activities; others use a "checklist." The IQ exercise sometimes takes 10 to 15 minutes, and unqualified people may perform it because it is a simple matter of "checkmarking" in the squares. This is particularly the reality whenever a change is done to the equipment, for which a typical question is, "Do we have to repeat the entire IQ again?" The reality is that too much attention is paid to paperwork, and this may preclude the firm from actual compliance and also from benefiting from the technical results and real benefit the IQ is supposed to bring. The best way to institute a sound IQ procedure is to prepare a guideline showing the typical features to be examined in any piece of

---

19. As an example, during the PQ runs at a supplier's plant, plastic parts being pressure pressed together tended to show a wider gap between the parts being assembled. Every time the neighbor assembly shop activated the pressurized equipment, the supplier's assembly line was affected. Nobody had realized that the hydraulic pressure source was common to all areas of the floor. An ancillary piece of equipment was missing—a pressure regulator.

20. Consider the case in which nobody checked on this item before the actual day of installation. The equipment was wider than the room, and the walls had to be broken to be able to install the equipment.

equipment. From this guideline, a protocol for execution can be derived according to the specific needs of the specific equipment and circumstances.

■ *Each time the equipment is moved, we must perform a new IQ.* Well-prepared validation work will include a PFMEA for each piece of equipment used in the process. Ideally, the PFMEA would indicate the effects from moving the equipment. Whether the PFMEA was done or not, when already qualified equipment is moved, the key strategy is to analyze the answers to questions such as the following:

☐ What could have been altered by motion? For example, if the equipment is an analytical balance, it makes a lot of sense to recalibrate it, at a minimum. Also, the balance should be on a still surface, such as a marble table, and far from HVAC vents or other sources of vibration or motion.

☐ What is different in the new environment?

☐ How about utilities and layout in the new location?

Operational Qualification

The second part of Figure 3.4 starts with OQ. Like IQ, the OQ is equipment oriented.

Definitions of OQ   According to the GHTF, OQ is "establishing by objective evidence *process control limits* and *action levels* which result in product that meets *all predetermined requirements.*" An analysis of the key words (italicized) in this definition yields the following:

■ *Process control limits* really implies process parameters such as time, temperature, and pressure settings of the equipment.

■ *All predetermined requirements* refers not only to acceptance specs, but also to product and process stability and reliability, and design-intended use (the OQ has to be based upon the DHF). For example, can gamma radiation weaken a given device to a point of breakage during use? See that the OQ can produce some important design outputs.

■ *Action levels* can be seen as "action limits" in SPC. That is, instead of reacting when a data point is outside of three standard deviation limits (99.73% confidence), the reaction and adjustment should come at two standard deviations (95% confidence).

The practical definition of OQ is "to understand and know how to control the manufacturing process." That is, which factors x_i can be controlled to ensure that the output characteristics y_i are within a predetermined range of values? In other words, OQ is to know what is going on inside the black box called process.

Process Characterization OQ includes process characterization. Process and product quality are more achievable when the people involved in the process know the what, how, when, where, and why. Answering those questions characterizes the process. An example is determining if the equipment can sustain a given range of parameters x_i for a specific amount of time or cycles.[21] For example:

- Can the equipment components wear out (are there replaceable parts)?

- Can the equipment maintain temperature, pressure, or other factors, or does it need periodic adjustments?

As part of the process characterization, different statistical tools will be used, such as screening and modeling DOE. Findings during the process characterization may present opportunities for:

- reduced downtime, scrap, cycle time, reworks, retests, or process optimization by elimination of unnecessary steps and redundant testing, and reduction of processing time and unnecessary sampling.

The process characterization phase of the validation effort is where the greatest economical and regulatory benefit to the company can be made.

Worst-Case Analysis Worst-case analysis (WCA) is one of the most confusing terms in the guidelines, especially the way it is applied. It is also known as worst-case conditions or worst-case test. Worst-case analysis implies process modeling, and an understanding of WCA requires an understanding of experimental design, or design of experiments. DOE is a statistical and scientific methodology used to create empirical models that describe the behavior of a process by establishing a relationship between input or independent variables x_i and output or a dependent variable y.

For example, Figure 3.5 shows three input variables, A, B, and C. The "+" means the highest experimental setting, and the "−" means the lowest experimental setting. By means of multiple linear regression and analysis of variance (ANOVA), a prediction model can be defined. Suppose that A = speed, B = temperature, and C = pressure. Then a typical multiple regression model is

$$Y = aA + bB + cC + e$$

where e is the experimental error. If the model is

$$Y = .02A + 30B + .5C$$

---

21. This is a very traditional concept about OQ. Specifically, it is very well known as such in pharmaceutical production. Some firms still do this activity as part of the OQ; others do it as part of the IQ.

RUN	A	B	C	OUTPUT Y
1	+	+	+	Y1
2	+	+	-	Y2
3	+	-	+	Y3
4	+	-	-	Y4
5	-	+	+	Y5
6	-	+	-	Y6
7	-	-	+	Y7
8	-	-	-	Y8

FIGURE 3.5 Experimental matrix.

what does it say? Just by the weight of the coefficients, temperature (*B*) is the most influential process parameter, whereas speed (*A*) basically has a minimal effect on the output variable *Y*. You can obtain the same nominal or target *Y* with different combinations of *A*, *B*, and *C*, not only to model worst-case conditions, but to obtain optimum settings. In this case, the prediction model is composed of only positive linear terms, so the worst-case condition is truly when the input variables or process parameters are either all at "+" or all at "−". In fact, the experimental matrix in Figure 3.5 shows that the worst-case conditions were included in the DOE experimental settings. However, if the model is

$$Y = .02A - 30B + .05C$$

the worst-case condition is different. This brings up a very important collorary in terms of worst-case conditions:

> Collorary 1: Worst-case conditions are defined in terms of *the output variable, not the inputs or process parameters.*

Thus, by using DOE and other statistical tools, the validation specialist will be looking for input effects, or process parameter settings, on the output variables.

The following are myths about worst-case analysis:

- *WCA means to manufacture three lots of the smallest and the largest products if the process is aimed at a family of products.*

- *WCA means to run the process with all parameters or controls at "high" and then at "low" settings.*

- *WCA is required in the PQ. It requires three lots at high and three lots at low.*

Besides compliance with FDA, what can be gained from WCA? Knowledge gained by challenging worst-case conditions may present opportunities for the following:

- Optimization of process parameters (e.g., speed, less costly settings, faster setup)

- Robustness of process

- Lot-to-lot consistency (reproducibility)

- Within-lot consistency (piece-to-piece reduced variance or repeatability)

- Fewer surprises for management, and better decision making, especially when trying to justify process deviations

- More objective evidence to include in the PFMEA and in the process control plan.

In summary, the worst-case analysis should be done as part of the OQ, not the PQ. The best tools are those related to DOE.

Short-Term Stability and Capability   As part of the OQ, at least 30 to 50 discrete parts should be made to evaluate stability and capability. The short-term stability of a process can be assessed with control charts and the short-term capability via potential process capability ratios such as the Cpk. It is said to be short-term because the design and development team and/or the validation team may not be able to assess the variance component of a signal caused by:

- shift to shift;

- raw material lot to raw material lot, vendor to vendor;

- air pressure/voltage surges;

- tool wear; or

- set up to set up.

Subsystem or Component-Level Reliability Analysis   Reliability testing is typically performed during design verification. However, this testing is usually done with laboratory-made prototypes or engineering pilot manufacturing runs, which may require more monitoring and "babysitting" than actual production runs. During process characterization, many worst-case potential components or subsystems can be produced. This is a golden opportunity to test the design for reliability. If the reliability is poorer than before, it could trigger an investigation about the effects of the manufacturing process. For example, in mechanical parts, designers use "stacking tolerances" to define component tolerances. By combining worst-case components (e.g., maximum and minimum interference between two components that form a subassembly), the design can be exposed to maximum challenge reliability testing.

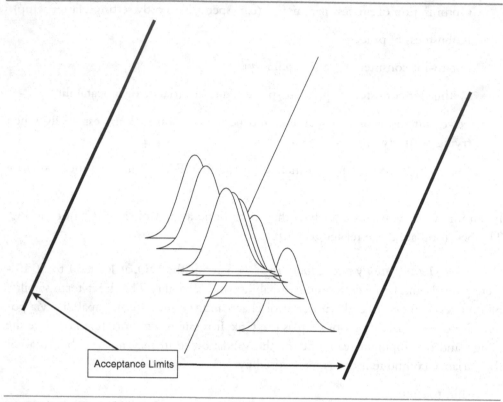

Acceptance Limits

FIGURE 3.6   Typical long-term process behavior.

Process Controls   From the experimental design results, the validation team can establish a relationship between product quality and reliability and process control parameters. For example, in sonic welding of plastic components, it is known that energy delivered by the horn onto the parts is potentially directly correlated with the strength of the welded bond. However, characterization via DOE is necessary because differences in shape, raw materials, and other variables will never guarantee such a perfect relationship. The necessity for validating sonic welding is that the testing to verify sonic weld strength is destructive. In summary, a good characterization of this process could lead the validation team to establish "energy delivered" as a parameter to control the process.

Process Stability and Capability   Process stability and capability is said to be short-term because not all the extreme cases may have been seen. Although Figure 3.2 shows the ideal behavior of the process, reality might look more like Figure 3.6. Figure 3.7 shows the short-term and long-term variance components.

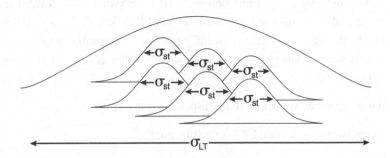

FIGURE 3.7 Long-term variability.

Multiple Processes or Manufacturing Steps The term "process validation" is confusing when multiple processes are involved. Some consultants have tried to differentiate processes from manufacturing steps. For a metallic part that is made by stamping and then goes to tumbling before assembly, either three processes or three manufacturing steps are all that are needed. Each one may require IQ/OQ and PQ. Usually, the IQ and OQ are done separately. Then the PQ is done for the "entire process." The concern here is depicted by Figure 3.8. What if the extremes of the manufacturing steps are combined? This is something that can be planned per the OQ protocols before moving onto PQ.

Performance Qualification

The definition of performance qualification according to the GHTF is "establishing by objective evidence that the process, under *anticipated conditions*, consistently produces a product which meets all predetermined requirements" (GHTF 1999b).

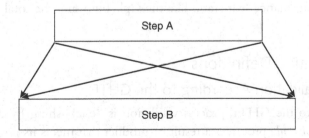

Can the product made at low in step A be affected by low or high in step B?

FIGURE 3.8 Interaction between manufacturing steps.

Anticipated conditions should not be confused with worst-case conditions. This is not the time to make worst-case analysis. The GHTF guidance is clear when it states "to demonstrate the process will consistently produce acceptable product under *normal operating conditions.*" This guidance talks about "challenges to the process." Examples of these challenges are:

- Normal equipment wear out (e.g., disposable parts)

- In-process adjustments per OQ recommendations

- Change in shifts

- Largest possible batch size

- Multiple raw material lots

- All possible sources of variance that can produce behavior, such as the one seen in Figures 3.7 and 3.8.

The parameter settings and process controls are supposed to be those defined and justified during OQ. The DMR is supposed to state such settings and process controls. Long-term stability should be assessed here. During the OQ, all process parameters were varied as part of the DOE and process characterization. In the OQ, these were elements of signal or variability. During the PQ, others that have not been accounted for may be seen. This is why we talk about short-term variability for the OQ and long-term for the PQ. Mathematically, this phenomenon is shown in the equation

$$\sigma^2_{LT} = \sum_{i=1}^{n} \sigma^2_{ST_i}$$

The practical definition of PQ is "to make validation runs at nominal settings, allowing all possible sources of variance to influence the process." This means the entire process, including packaging and sterilization. The PQ should include evaluation of the process performance (e.g., ratio like the Cpk, but using the total variance of the process).

Process Validation Definitions

Process Validation According to the GHTF

According to the GHTF, process validation is "establishing by *objective evidence* that a process *consistently* produces a result or product that meets its *predetermined specifications.*" An analysis of the key words (in italics) in this definition follows:

- *Objective evidence* means well-applied science and statistical methods that can convince anybody that the conclusions are correct. In process validation, it is important to remember that medical devices are based on science. Science has everything to do with the effectiveness of a medical device. A sound method of development and design of a product and its manufacturing process signals well-applied scientific principles. The word "objective" here means that data are more powerful than any expert's opinion.[22] "Well-applied statistics" are cited because the only way to approach variation is by mathematical formulas. Because the evaluation of process variation may include sampling, the correct use of statistical methods (21 CFR 820.250) is essential.

■ *Consistently* is as described in Figures 3.2 and 3.3. In terms of six-sigma programs and other total quality initiatives and standards, this is actually process stability.

■ *Predetermined specifications* mean that before executing the process, the design and development team have specified product and process characteristics that, when measured, must meet some criteria. The ability of a manufacturing process to meet the predetermined specifications is typically evaluated via process capability ratios. It is said to be predetermined because it makes no sense to let the process dictate the acceptance limits. This is why, among other reasons, the first step in process validation is to understand the design and the specifications to be met, as shown in the flowchart of Figure 3.4.

Figure 3.9 presents the GHTF process validation decision flowchart. Notice that process control and risk analyses are required. This strengthens even more the position that the validation team must start its work by understanding the DHF. Without risk analysis, DFMEA, and PFMEA, how will anybody be able to assess risk? Without a corporate procedure for risk analysis and management, how can R&D and manufacturing be consistent about risk levels? Another interesting point in Figure 3.9 is box H—it calls for redesign of the product and/or process! This also strengthens the position that process validation is under the umbrella of design transfer and that there is a role for factory personnel in design control matters. The GHTF has complemented 21 CFR 820.75 by providing industry with this guidance decision-making tool.

Box E in Figure 3.9 is a controversial decision. In conversations with FDA inspectors and former inspectors, they have communicated that this decision is not acceptable. The authors want to remind the readers that the 1987 Guideline on General Principles of Process Validation remains as "the guideline" (FDA 1987).

Process Validation According to the FDA

According to 21 CFR 820.75, "Where the results of a process cannot be *fully verified* by subsequent inspection and test, the process shall be validated with a high de-

---

22. A company "expert" may state a hypothesis or "opinion," which then becomes the basis for a spec or an acceptance limit. This is precisely what must be avoided. If the phenomenon is not a natural law (e.g., $V = IR$, Ohm's law), then the design and development or validation team must design the empirical test or experiment to achieve a level of objectivism.

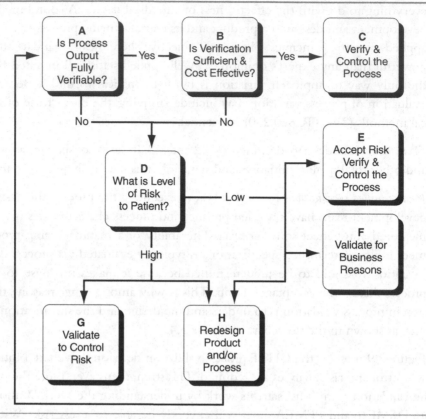

FIGURE 3.9 GHTF process validation decision flowchart.

gree of assurance and approved according to *established procedures*." Thus, we could say that verification allows us to waive the validation requirement, but, as will be shown shortly, this is not the case. An examination of the key words (in italics) in this definition helps to explain why.

- *Verification* is defined by the GHTF as confirmation by examination and provision of objective evidence that the specified requirements have been fulfilled.

- *Fully verified.* First, note that "100% inspection" can lead to full verification, but waiving the requirement for process validation is not automatic. The design and development and the process validation teams must still go through some of the work in Figure 3.4. The manufacturing process must be characterized if either team wants to waive the validation. Remember that "100%" implies all units, all characteristics, and all conditions. One example of a process that can be fully verified is the assembly of printed circuit boards (PCB). By providing an in-circuit test (ICT) combined with burn-in testing, essentially all relevant electrical and/or electronic parameters can be automatically tested unit by unit. Of course, the equipment and the software that run the ICT would still need to be validated.

- *Established procedures.* Process development, like product design, is iterative. When making the validation runs (e.g., the PQ, as explained earlier in this chapter), the process is supposed to be fully characterized. Thus, whether the validation team is going to validate or just to verify, many of the activities depicted in Figure 3.4 still must be performed.

Practical Definition of Process Validation

The practical definition of process validation is the same as that for design transfer—see that consistency is achieved via repeatability and reproducibility. Reproducibility, in mathematical terms, is trying to obtain the same probability distribution lot after lot, as shown in Figure 3.2. Repeatability is minimizing the spread or standard deviation from item to item within a lot, as shown in Figures 3.2 and 3.3. If the DMR does not correctly translate device design into production specifications, the process validation may fail, and thus design transfer also may fail.

A validated process is a predictable process. The manufacturing team can rest assured that by setting the equipment according to predetermined specifications and by using the specified materials under the specified conditions, the output product will meet the acceptance specifications. This implies that the acceptance specifications will guarantee that the design will meet the intended use of the device. Thus we see that the expected results of process validation are the same as design transfer.

Other Regulation Definitions

21 CFR 820.75(b) states that "each manufacturer shall establish and maintain procedures for *monitoring* and control of *process parameters* for validated processes to ensure that the specified requirements continue to be met." This message is consistent with continuous improvement. Thus the validation never ends. An examination of the key words (italics) yields the following:

- *Monitoring* does not have to be continuous; if done periodically, it should be justified statistically and its risks evaluated (e.g., by using a PFMEA and statistical analysis).

- *Process parameters* are parameters such as speed, temperature, and so on. A typical mistake is to rely on acceptance data.

21 CFR 820.75(b)(1) states, "each manufacturer shall *ensure* that validated processes are performed by *qualified individual(s)*." Key words here are "ensure" and "qualified individuals":

- *Ensure* is not equal to assure. "Assure" refers to affirmation, assertion, or declaration, whereas "ensure" implies possessiveness, certainty, or certitude.

- *A qualified individual* is an individual who has the training requirements for the job function as well as actual training evidence. All those involved in executing the validation runs must understand the steps and the rationale behind them. Include all applicable documentation. An implication of subsection (b)(1) is that if during the actual validation runs (PQ), all the quality systems at your company are being challenged, a validation failure because "somebody needs retraining" disqualifies your training program (21 CFR 820.25).

21 CFR 820.75(b)(2) states that "for validated processes, the monitoring and control methods and data, the date performed, and, where appropriate, the individual(s) performing the process or the major equipment used shall be documented." This reinforces some of the elements to include and save in the DHR.

21 CFR 820.75(c) states, "when changes or process deviations occur, the manufacturer shall review and evaluate the process and perform *revalidation* where appropriate. These activities shall be documented." If appropriate monitoring systems are in place and the process can be controlled, a continuous process validation (e.g., *revalidation*) is going on. If, conversely, process changes occur, new revalidation may have to take place. But what is revalidation? This is *not* running three more lots and seeing what is happening. Revalidation implies a review of the process shown in Figure 3.4. Specifically, we may think that, in general, new variables can be introduced that affect the process, such as new conditions or a new raw material vendor. Process recharacterization may need to take place.

Myths About Process Validation

The following myths point out misconceptions about process validation:

- *Running three lots following approved documents qualifies as process validation.* The famous "three lots" practice is not stated in the regulation nor in any guidelines.[23] Also, process validation is not a matter of running three or *x* number of lots; it is a matter of knowing what is happening and achieving the desired expectations. Guidelines suggest "appropriate" sample size or "enough times to assure that the results are meaningful and consistent." As an example, a contract manufacturer

---

23. Commentary no. 85 from the 21 CFR 820 Preamble (October 7, 1996) states, "The requirement for testing from the first three production lots or batches has been deleted. While FDA believes that three production runs during process validation (process validation may be initiated before or during design transfer) is the accepted standard, FDA recognizes that all processes may not be defined in terms of lots or batches. The number three is, however, currently considered to be the acceptable standard. Therefore, although the number requirement is deleted, FDA expects validation to be carried out properly in accordance with accepted standards, and will inspect for compliance accordingly."

was showing identical results on three validation lots. Similar products had never been so consistent from lot to lot. Upon review of the routers and batch records (DHR), it was discovered that the three lots were really the same manufacturing lot that had been divided into three shipping lots. In essence, the aim for challenging reproducibility would never be achieved this way.

Process validation practitioners should always keep in mind that using multiple lots or batches is really aimed at simulating long-term process performance. All the quality systems and other process controls should be challenged and should show repeatability and reproducibility. For example, a manufacturer may decide that the highest source of variability or inconsistency in the manufacturing process depends on the three different shifts of the workforce. Thus the manufacturer may decide to ask each of the three shifts (reproducibility) to built at least two batches (repeatability) for a total of six batches. Another manufacturer may decide that in a highly automated process, the highest source of variability or inconsistency in the manufacturing process depends on raw material lots. Thus that manufacturer may decide to use a different raw material lot in each of his three validation runs.

- *All products have to be validated.* There is no such a thing as "product" validation. Processes and designs are to be validated. When a new product is to be run in validated equipment, a need arises for certain minimum evaluation such as process characterization. The results of such process characterization should be the criteria for deciding the next steps (e.g., further characterization, WCA, PQ).

- *Worst-case analysis or worst-case testing must be run during the PQ.*[24] Most of the activities for WCA are to be done during the OQ.[25] During the PQ, the process is challenged by simulating conditions (frequent and infrequent) that will be encountered during actual manufacturing. The key to understanding what to do really comes from realizing that the DMR documents[26] are supposed to be updated by the manufacturing and/or process validation teams during the process characterization phase of the OQ. The new DMR documents should include such concepts as action limits that should be challenged during the PQ. Other challenges can include changing disposable parts of the equipment in the middle of a run, combining different raw material lots in the middle of a run, including all shifts of personnel, and so on. In fact, if "rework" or "reprocessing" will be allowed, this should also be validated (design and process). As an example, certain plastic components on medical devices could lose strength and other physical

---

24. This misconception is so significant that it has been published in papers and journals. Also, companies have wasted significant amounts of resources and scrapped products because processes were not characterized before running worst-case lots or batches.

25. As part of the process characterization (see Figure 3.4).

26. For example, manufacturing or work instructions.

properties with the accumulation of gamma radiation. Thus, as part of the reliability assessment (design verification[27]) and process validation, instead of passing the product once through gamma radiation, the product is passed twice, simulating a rework[28] activity.

- *If a process is validated, it should not fail to meet the acceptance specifications for a given batch.* 21 CFR 820.75(b) states, "Each manufacturer shall establish and maintain procedures for monitoring and control of processes parameters for *validated processes to ensure that the specified requirements continue to be met.*" What this means is that the validation of the process never ends. A means of surveillance (i.e., monitoring) and corrective action or adjustment (i.e., control) is part of the umbrella concept called process validation. That is, processes can suffer deviations, and the manufacturer must ensure that even after successful validation, effective and valid process controls are in place to detect such potential deviations. These process controls should be identified during the process characterization done as part of the OQ.

- *The process validation team does not need to have access to the DHF.* During the activities portrayed in Figure 3.4, a need typically exists to understand the rationale behind design specifications and tolerances. While trying to achieve stability and process capability, situations can arise in which the validation team may ask questions such as, "Can we change the spec?" or "Can we widen the tolerance?" Answering these questions will be very difficult unless the R&D team is involved in the validation of the manufacturing process or the DHF is available to the validation team.

- *There is no need to validate test methods because we follow USP procedures.* Using a USP procedure is good, but we have to widen the view. How do we know that we are measuring with the right procedure? Do we comply with the necessary test method requirements? What about sampling and handling the sample? What about qualifying QC technicians?

- *GR&R is not needed because all measurement equipment have been calibrated or are in the calibration program.* Although this is important, the purpose of GR&R is to assess the total error of measurement, or measurement capability analysis (MCA). For example, a piece of well-calibrated equipment may be susceptible to variability among QC technicians. Thus, its MCA may not have the resolution

---

27. The position of the FDA is to show evidence that rework does not cause an adverse effect on the product, and this is to be documented in the DHF. In fact, this is a typical entry to the PFMEA.

28. The FDA has adopted the word "rework" instead of "reprocessing" to be consistent with ISO 8402:1994.

needed. Usually, calibration has nothing to do with repeatability and reproducibility.

■ *There are no problems with the raw material or the parts supplier because they are certified.* Like process validation, supplier certification[29] is another misused and misunderstood concept. Certification usually means conformance to a quality system standard such as ISO 9002, but would this guarantee conformance to your specifications? The validation team should evaluate risks and include the supplier's evaluation for the specific parts or raw material of its project as part of the validation plan. For example, in injection molding of plastic or metallic parts, each new mold and/or cavity needs validation, no matter the certification status of the supplier.

■ *If the change is only a scale-up, we do not need to revalidate.* If you are using new manufacturing equipment, you need to run IQ/OQ at a minimum. You possibly may not need to do all the characterization activities depicted in Figure 3.4, but some level of understanding is required. This is to be defined in a validation plan.

■ *Transferring production to the big factory in country X should be no problem because in R&D we had a pilot plant.* Pilot plants have to be formally validated just as any other manufacturing plant no matter the size. Pilot plants are typically under the fatherhood of R&D. Many changes typically take place, even after validation. However, because "design experts" are readily available to decide whether manufacturing changes impact intended use and risk levels, no big operational issues are visible to management. The fallacy comes into place when the production is transferred to a new place. Sometimes the machines are new; also there is no such a thing as "identical machines." At the new place, IQ, OQ, and PQ need to take place. Management cannot take the approach of merely making "confirmation runs" because the process "is the same as the pilot plant." Without proper process characterization, the process at the new place can be quite different from the pilot plant. For example, if the process is susceptible to environmental conditions, then a transfer from Chicago (cold and dry) to Florida (hot and humid) can be quite different.

Should We Just Verify?

Another way of looking for an answer to the issue of whether we should verify is to look at Figure 3.4 and seek answers to the following questions:

---

29. We have known cases in which a QA manager has a goal of "certifying two suppliers" in a year. But should the goal be to certify the supplier or to get good parts from the supplier? Maybe this QA manager has to train the supplier for many months before he can even consider "certification."

- Can we afford *not* to validate?

- Can we afford *not to know what is happening with the process*?

- Can we afford not to know the critical process parameters?

- Can we afford not to know the effect of those parameters on the quality attributes and reliability of the product?

Summary: Can We Sell These Lots?

We first need to validate design. In the 1987 process validation guidelines from the FDA, the term "product qualification, or product performance qualification" was aimed at ensuring that the manufacturing process had not adversely affected the integrity of the product (FDA 1987). *This is basically the concept of design validation.* Figure 3.4 closes the loop and comes back again to design transfer and design control. The second part of Figure 3.4 calls for closing the loop and includes the next to last step as design validation or product qualification. This is a gap in the GHTF guidance for process validation (see Table 3.2). We kept the term "product performance qualification" in the table because some firms may argue that once a design and a process is fully validated, future revalidations (e.g., due to product transfer) may not need as rigorous a design validation as when the product was first introduced to the market. This is a valid point. In our view, we could still keep the term "product qualification or product performance qualification" for this reason. The main issue we see in today's industry is the lack of awareness about the need to ensure that the device works as intended.

TABLE 3.2 Understanding Process Validation Guidances

Source	Phase I	Phase II	Phase III	Phase IV
This book	Enhanced IQ plus error of measurement and test methods	Enhanced OQ plus linkage to product reliability	Performance qualification (PQ—aimed at process) linkage to product reliability	Design validation and/or product performance qualification
1987 FDA Guidance (FDA 1987)	IQ	OQ	Process performance qualification (PQ)	Product performance qualification
1999 GHTF Guidance (GHTF 1999b)	Enhanced IQ	Enhanced OQ	Performance qualification (PQ—aimed at process)	

Further Reading

DeSain, Carol, and Charmaine Vercimak Sutton. 1994. *Validation for Medical Device and Diagnostic Manufacturers*. Buffalo Grove, Ill.: Interpharm Press.

Schmidt, Stephen R., Mark J. Kiemele, and Ronald J. Berdine. 1996. *Knowledge Based Management: Unleashing the Power of Quality Improvement*. Colorado Springs, Colo.: Air Force Academy Press.

Schmidt, Stephen R., and Robert G. Launsby. 1994. *Understanding Industrial Designed Experiments*. Colorado Springs, Colo.: Air Force Academy Press.

Trautman, Kimberly. 1997. *The FDA and Worldwide Quality System Requirements: Guidebook for Medical Devices*. Milwaukee: ASQ Quality Press.

CHAPTER FOUR

Quality System for Design Control

This chapter presents the basic and practical elements necessary to build and maintain a quality system that ensures compliance with the design control requirements. Practical examples of key procedures are presented.

Product Design and Development Process

To design a product and develop its process, the team responsible usually proceeds in steps or phases. Typically, R&D organizations have their own jargon by which they describe such a process, such as concept feasibility, prototype testing, prototype refinement, process development, process characterization, process qualification, engineering pilot runs, product and/or process adjustments or refinements, qualification runs, and design validation. The sequence, name, and meaning of the phases vary from firm to firm. However, in practical terms, they all know that design and development is an iterative process in which the product is tested several times to demonstrate that it meets a defined intended use. Furthermore, firms are concerned about meeting a product's intended use under certain conditions and for a stated amount of time or cycles. In other words, a firm's main concern centers around the following questions: What is the device's reliability? How do we show it works? How shall we document the proof?

Procedures are aimed at providing guidelines and checking for completeness of the design and development activities. Simultaneously, procedures provide for compliance to the regulation, and they standardize the activities and design outputs. The larger the company, the greater is the need for standardization.

The typical ISO 9000 quality system divides the documentation system into quality manual, procedures,[1] work instructions, and records. (The original four levels

---

1. Also known as Standard Operating Procedures (SOP), or in some companies as just Operating Procedures (OP), or in others as Corporate Procedures (CP).

were changed to three levels by combining the last two in 1994.) We can think that the DMR is composed of specific work instructions to the process and the designed product; whereas records will end up being part of the DHF and DHR.

Implementation of design control requires a revision to all three levels of documentation. The quality manual should adopt policies for design control and process validation among all other quality systems. The policies stated in the quality manual are the evidence of management responsibility (21 CFR 820.20). Table 4.1 presents an example in which all three levels of documentation are used to define a design control quality system.

The Design History File

The Design History File (DHF) contains documents such as the design plan and input requirements, preliminary input specs, validation data, and preliminary versions of key DMR documents. These are needed to show that plans were created, they were followed, and specifications were met. The DHF does not have to contain all design documents or to contain the DMR; however, it will contain historical versions of key DMR documents that show how the design evolved (e.g., design specification and design drawings).

Does the DHF have value for the manufacturer? Yes, when problems occur during redesign and for new designs, the DHF has the "institutional" memory of previous design activities. For example, various versions of the risk analysis and FMEA may indicate why the design iterations took place and what was improved from iteration to iteration. The DHF also contains valuable verification and validation protocols that are not in the DMR. This information may be very valuable by pointing to the correct direction to solve a problem; or, most important, preventing the manufacturer from repeating a design that has already been tried and found to be useless.[2]

Typical documents that may be included or referenced in a DHF[3] are the following:

- Records of customer inputs
- Design and development plan(s), including packaging and sterilization
- Quality and reliability plan(s)
- Regulatory plan or strategy
- Design review meeting information and notes

---

2. This is a common practice in some high-tech companies.

3. Remember that the DHF is not a binder, but a concept.

TABLE 4.1 Example of a Design Control Quality System

Waterfall Model (GHTF 1999)	Quality Manual Examples of Policy Statements	Standard Operating Procedure(s)[a]	Work Instructions	Records
User Needs	Every product to be developed shall include user need data. These data are to be collected by marketing, business development, and R&D. Clinical and preclinical evaluations shall be considered accordingly.	General procedure stating how to prepare questionnaires, surveys, and so on. Example of related SOPs: • Statistical methods • QFD • Company personnel behavior when dealing with consultants and potential customers	Work instruction stating a standard way of defining the product requirements or product criteria for design and development.	Product requirements or product criteria for design and development. Save all records and data gathered. Identify all information with dates and names.
Design and Development Planning	Once there is a commitment from the company board of directors to go ahead with a project for new product design, the first step to be taken by the R&D project leader is to prepare a DADP.	General procedure stating all typical details needed in a DADP.	Work instruction detailing all typical activities at each phase of design and development that at the same time meet the design control requirements. This work instruction basically creates the DADP.	A DADP with all the revisions.
Design Input	Once the user's needs data are gathered, analysis by marketing, business development, and R&D will screen the important information according to company goals and the related regulations in the industry. Such filtered data will become the design input.[b]	Procedure stating the process by which design inputs are converted into design outputs. Examples of related SOPs: • Procedures for field quality data tracking • Procedures for QFD, marketing data assessment, and documentation	Work instruction to create: • Design specification • Quality/reliability plan • Marketing plan • Regulatory plan or strategy • Review and analysis summary report of field quality records[c]	Design specification. Quality/reliability plan. Marketing plan. Regulatory plan or strategy.d Field quality requirements summary.

continued next page

TABLE 4.1 *continued*

Waterfall Model (GHTF 1999)	Quality Manual Examples of Policy Statements	Standard Operating Procedure(s)[a]	Work Instructions	Records
Design Process	Every project will follow the company's defined phases of design.[e]	General procedure describing the phases of design and development and the responsibilities of those involved. Procedures for designing manufacturing equipment and tooling, sterilization methods, reliability testing.	Work instruction detailing all typical activities of each phase of design and development that at the same time meet the design control requirements. This work instruction basically creates the DADP.[f]	All deliverables defined in the DADP.
Design Output	Design outputs are considered acceptable if there is documented evidence that they meet design inputs. Design outputs include the DMR and the completion of activities in the DADP.	General procedures for: • Protocol creation and documentation • Risk analysis and management • Design verification • Design reviews • Design and process validation	Work instructions for: • Risk analysis and management • Design verification • Design reviews • Design and process validation	Project-specific records of: • Risk analysis and management • Design verification • Design reviews • Design and process validation
Design Review	The design and development activities must be reviewed at appropriate stages of the project. Such reviews will be stated in the DADP. A review leader will appoint independent reviewers according to their areas of expertise.	General procedure stating how to plan, coordinate, hold, document, follow up the issues, and close a design review.	Work instructions to generate a design review roster, agenda, list of deliverables,[g] pending issues, ways to follow up with those pending issues, and how to document closing such issues.	Records for each design review showing evidence that it really took place, that it was per the DADP, that the open issues were closed, and that the project was given approval to proceed to the next phase.

Design Verification	The design and development team will demonstrate that design outputs met design inputs. Such evidence will be documented in the DHF.	Same as design output and design reviews. Other procedures are: • Design reliability evaluation and testing	Same as design output and design reviews. Other work instructions: • Reliability methods for testing and analysis.[h]	Same as design output and design reviews.
Design Validation	Design validation must ensure that customer needs are actually met. Design validation will be done with initial production units such as units made during process validation. Deviations to this policy are allowed as long as they are documented in the design validation protocol and approved per the design review. A systems reliability test or equivalent is part of design validation.	Same as design output and design verification.	Same as design output and design verification.	Same as design output and design verification.
Design Transfer	Design transfer is an integral part of the responsibilities of the design and development team. Process validation takes place as part of design transfer.	Same as design output, verification, and validation. Additional procedures are: • IQ/OQ/PQ • GR&R • Test methods validation • Calibration • Associate training • Release of DMR from R&D to operations • BOM/routers preparation	Same as design output, verification, and validation. Additional work instructions are: • IQ/OQ/PQ • GR&R • Test methods validation • Calibration • Associate training • Release of DMR from R&D to operations • BOM/routers preparation	Same as design output, verification, and validation. Additional records are: • IQ/OQ/PQ • GR&R • Test methods validation • Calibration • Associate training • Release of DMR from R&D to operations • BOM/routers preparation • DHR

continued next page

TABLE 4.1 *continued*

Waterfall Model (GHTF 1999)	Quality Manual Examples of Policy Statements	Standard Operating Procedure(s)[a]	Work Instructions	Records
Design Changes	There are two kinds of design changes: Design changes during the phases of design and development must be documented and controlled in such a way that the remaining steps in the project are adjusted accordingly. These design changes are to be approved by the design and development team and their reviewers. Design changes after the product has been released to the market necessitate a "design change plan" to be considered according to the risk levels and/or effects on the intended use of the device. Such design change plan shall include design validation. If only design verification is needed, a justification shall be stated and approved.	General procedures for: • DMR change control during design and development and after product has been released to the market • Design change control	Design change control. Change order[i] generation and approval.	Design change control. Change order.
DHF	The DHF will contain all quality records per the DADP and, when applicable, per the design change plan.	General procedure indicating how to organize and file DHF records.[j]	Work instructions indicating how to organize and file the DHF records.	The DHF itself.

A controlled copy of the DHF must remain with the manufacturing operations quality assurance group because changes or deviations may necessitate review of such records.

The DHF will remain in record per company policy regarding document retention.

- Sketches and/or drawings (DMR)

- Procedures (e.g., for medical capital equipment or for performing preventive maintenance and calibration by the customer)

- Photos

- Engineering notebooks

- Component and/or raw material qualification information

- Biocompatibility (verification) protocols and data

- Verification protocols and data for evaluating prototypes

- Validation protocols (e.g., design and manufacturing process) and data for initial finished devices

- Contractor/consultant information

- Parts of any design output or DMR documents that show plans were followed

- Parts of any design output or DMR documents that show specifications were met

Design and Development Plan Quality System

A procedure to develop design and development plans (DADPs) is a must for a medical device company. The DADP is the main source of control for the design and development team as well as the main reference point to design control auditors and design reviewers. Once the concept of what is to be designed is clear and a formal project is supplied with resources, the DADP is typically the first formal design control element or deliverable to be built. We say "first formal" because, in real life, at this stage a lot of design input has been gathered and tools of quality aimed at customer satisfaction (e.g., QFD, SWOT,[4] industry structure analysis) have possibly already been applied. This plan not only lists the design and development activities and responsibilities, it also links all the required company quality systems into an integrated network of events.

Prerequirements

Before developing a DADP procedure, the firm should look at its way of developing new products. This can be considered as the phases of new product design and devel-

---

4. SWOT is a marketing tool used to evaluate a company's strengths (S) and weaknesses (W) as well as the opportunities (O) and threats (T) it faces in the market or industry environment (Donnelly 2000).

opment. Although not a requirement of the regulation, design and development teams work better when guidelines are provided and such guidelines are flexible enough to allow for "creativity." Such guidelines should map the company's way of designing within the design control requirements. For example, the June 29, 1999, GHTF "Design Control Guidance for Medical Device Manufacturers" presents the "waterfall design process." To device design engineers and R&D managers, this model is not very useful; however, to quality systems engineers and quality managers, the model is very clear. The DADP procedure can interlock the waterfall model (design control requirements) with the firm's phases of new product design and development. An example is provided in Table 4.2.

Elements of the Design and Development Planning Procedure

If a general procedure is not in place for definitions and acronyms for all company quality systems in general, the DADP procedure should start defining those that are applicable. Company jargon such as what is meant by "pilot runs," "qualification runs," and so on should be included. In real life, large companies with a homogeneous product line develop work instructions, which serve as a DADP template, that contain a checklist of many of the typical deliverables for each phase of the project. The following is a list of DADP elements:

- Front page
    - Title and brief description of the project
    - Approver's name, title, and specific project responsibility or role
- Define activities according to the design and development phases (see example in Table 4.2) and assign responsibilities.
    - The design and development phases must be defined. For example, what is the concept feasibility? What is the objective of this phase?
    - Define the technical interfaces and their role.
    - Include responsibility for organizing and filing the DHF.
- Define how many design reviews will take place and the deliverables to be discussed in each review.
    - Generate a roster for the design and development team members that defines deliverables for each phase of the project and identifies the reviewers.
- Establish a relationship between the design and development phases and the design control requirements (see example in Table 4.2).
- Establish special rules and activities (e.g., an IDE that requires a special limited release to perform clinical evaluations).

TABLE 4.2 Example of Phases of a Design and Development Project and Design Controls

Waterfall Model Elements	Concept Feasibility[a] (Predesign Control Phase)	Design Process (Design Control Starts Here)	Phases of a Design and Development Program				
			Prototype	Prototype Testing	MFG Process Development	Preproduction	Initial Full-Scale Production & Market Release
User Needs	Define user needs.						Design is validated against user needs.
Design and Development Planning	Needs the output from this phase as the basis for planning.	DADP is generated and approved.	DADP is updated.	DADP is updated.	DADP is updated.	DADP is updated.	DADP is updated for the last time.
Design Input	Can use customer data collected as records and input for analysis. One or more intended use(s) are considered.	Customer needs lead to a specific intended use. Design inputs are deemed to be "final" or "frozen."	Design outputs try to meet design inputs.	Testing aimed at showing that design outputs meet design inputs.	Manufacturing process is developed based on design inputs.	Process characterization units are used to evaluate compliance with design inputs.	
Design Process		Design inputs are translated into design outputs.	Design iterations and prototype build.	Evaluate prototype; redesign and test again. Start evaluation of suppliers.	Design and develop manufacturing equipment.	Manufacturing process changes and improvements	

Design Output	Initial DMR drafted documents. Design specification is the technical driver of the project.	DMR refined documents.	DMR further refinement.	DMR is available for pilot runs.	DMR is challenged and the tolerances and specs are assessed. Process capability and stability are assessed.	DMR is transferred to manufacturing.
Design Review	At the end of concept and feasibility, if project is approved. First design review, to present the project to reviewers and discuss preliminary DADP elements.	Second design review, to evaluate all plan elements and design inputs as well as first prototype(s): • Risk analysis • FMEAs • Quality and reliability plans • Material evaluation • Tooling requirements evaluation			Third design review, to evaluate all design outputs, reliability, safety, efficacy, and process capability.	Fourth design review, to evaluate final results from design and process validation. If all requirements are met, product is released to market.
Design Verification	Design verification is planned as part of the DADP.	Evaluate prototype against design input.	Evaluate design output against design input.	Evaluate design output against design input.	Evaluate design output against design input.	
Design Validation	Design validation is planned as part of the DADP.	Design validation plan is updated.	Design validation plan is updated.	Design validation plan is updated.	Design validation plan is updated.	Design validation is executed with initial production units.

continued next page

TABLE 4.2 *continued*

Waterfall Model Elements	Concept Feasibility[a] (Predesign Control Phase)	Phases of a Design and Development Program					
		Design Process (Design Control Starts Here)	Prototype	Prototype Testing	MFG Process Development	Preproduction	Initial Full-Scale Production & Market Release
Design Transfer		Design transfer is planned as part of the DADP.	Update transfer plan.	Update transfer plan.	Design transfer is started. Generate process validation plan.	Go through process validation flowchart (Chapter 3). Evaluate and update DMR according to IQ/OQ results.	Design transfer is completed if process and design validation are successful.
Design Changes	N/A	Per R&D change order[b] system	Per R&D change order system	Per R&D change order system	Per R&D change order system	Per R&D change order system	Per company's traditional change order system[c]
DHF	Customer needs are documented and filed (first design and development records).	The DADP with the customer needs become the first elements of the DHF.	DHF buildup.	DHF buildup.	DHF buildup.	DHF buildup.	DADP checklist is verified for completeness of all deliverables. Final configuration of the DHF triggers product release to the market.

[a] Although not required by the FDA, it is a good practice to start with reliability planning and at least a preliminary hazard analysis in this phase. The idea to start with these two elements of the DHF will come when design inputs start being conceived and the associated technologies start being evaluated.

[b] Each company shall have a procedure defining how changes to DHF and DMR elements are to be completed. The main reason this is special is that all these documents and records are still within R&D.

[c] Now the DMR is in the hands of operations. At the end of the preproduction phase, R&D is supposed to "release" the "preliminary" DMR to production operations.

Example of Phases of a Design and Development Project

The example in Table 4.2 is aimed at illustrating the interlock between the phases of a design and development project and the design control requirements. The definitions of such phases are as follows:

- *Concept feasibility:* This is a data-gathering and analysis stage. Business development and marketing work together to speculate on business and market potential opportunities. R&D enters this iterative process to assess available technology and/or define the kind of technological breakthrough needed (technical feasibility). Eventually, financial "pro forma" models will define the project's net present value and options. The FDA does not require design controls in this stage, but design control quality systems (e.g., risk analysis, QFD) would be very useful in this phase. Should top management approve the project, the data on customer needs gathered here become records for design input. Also, the technical evaluations are the initial input for the design process, specifically the design specification.

- *Design process:* In this phase, an official project is started. The DADP and the design inputs are supposed to be the first two elements of the DHF to be approved. The GHTF design control guidance of June 29, 1999, states that design control starts with the approval of design inputs (GHTF 1999a). To make the approval of design inputs formal and official, an entity must be created. In some companies, this is called "product criteria or product requirements"; in others, it is "product goals or design goals." This is done with other purposes: first is the need to condense and summarize all the design inputs in a standard fashion,[6] and second is the practical need to have a voice of the customer entity that triggers the design specification.

- *Prototype and prototype testing:* Most of the design iterations take place in these two phases. Companies must have procedures for protocol generation and documentation as well as procedures for DMR control while the prototype is still being built by R&D.

- *Manufacturing process development:* New machines or equipment must be designed, ordered, tested, and characterized. For example, in the case of injection molding, molds have to be designed and ordered.

- *Preproduction:* This is a pilot phase. Design transfer activities are started as well as all characterizations, engineering studies, IQ/OQ, GR&R, test method validations, and so forth. (These activities are addressed in Chapter 3.)

---

6. This is so that the design and development team can start working on generating the design specification.

■ *Initial full-scale production and market release:* This is the end of the design and development cycle. A lot of DHF and DMR elements are finalized in parallel fashion in this phase. PQ and design validation are the two most crucial outcomes in this phase.

Design Guidelines

Design guidelines should be established to guide key personnel on how specific aspects of the design and development project shall be executed. These will include value engineering, reliability and maintainability, configuration control, interchangeability requirements, functional trials, safety requirements, and so on. We emphasize again that the FDA does not require the industry to have design and development guidelines. However, our 30-plus years of combined experience tell us that these guidelines are crucial to nurture good design and development disciplines as well as the standardization of related activities and their documentation. Table 4.3 presents some examples.

Design History Matrix

The DHF will be composed of many sources of information. The interrelation among these sources of data will be obvious to some and not that obvious to others. Whenever a design and development team member wants to ensure that a customer need was correctly translated into design inputs, and that design input was then correctly

TABLE 4.3   Design Process Guidelines

Key Identifier	Your Company Response (Yes/No/Maybe)
Do you practice value engineering?	
Do you have guidelines for reliability and maintainability?	
Do you have guidelines for configuration control?	
Do you have guidelines for interchangeability requirements?	
Do you have guidelines for functional trials?	
Do you have guidelines for safety requirements?	
Do you have a list of approved signatories?	
Do you have guidelines for control of software?	
Do you have guidelines for corrective action procedures?	

translated to design output, and finally that design output was correctly transferred to manufacturing, the task will not be easy.

We therefore introduce the concept of a design history matrix to track design development and fulfillment.

The design history matrix is neither a new FDA requirement nor a new GHTF guideline. It is just a tool to monitor and ensure design control from the early to the final phases of the design and development process. The idea is to take a customer need as it came from the customer and show its entire trajectory throughout the design and development process. All pertaining records and documents will be listed, and thus traceability to the DHF and DMR is simple. Table 4.4 presents an example of a design history matrix.

Design Change Plan

As indicated in Chapter 2, there are two main circumstances in which design changes may have to be considered. The first is during the stages of design and development. The second is after the product has been launched into the market.

During the stages of design and development, the design change plan is really an update to the DADP and all other subplans such as: reliability and quality plan, testing plan, material qualification plan, supplier qualification/certification plan, and so on. Also, all "preliminary" DMR elements must be updated. The DADP can include something like a summary that evaluates all the consequences of such action.

In our view, the most critical design change occurs after the product has been released into the market. Any change(s) to either DMR, or the process or the design shall call for an evaluation of the potential impact on safety and performance of the device. The design change plan in this case should include the following minimum elements:

- Product(s) being changed

- What is being changed, for example:
  - Process parameter ranges or nominal values
  - Test method(s)
  - Physical characteristics of the device or a component
  - Storage conditions
  - Raw materials

- Motivation(s) for change(s)

- Which organization is responsible for initiating and completing the change(s)?

TABLE 4.4 Example of the Design History Matrix Using an Intravenous (IV) Tubing Set

Customer Needs and Wants—Source	Input for Design	Design Specification Requirement	Design Output (DMR)	Design Verification	Process Validation	Process Controls	Design Validation
Customer Survey #xxx	Shelf life	5 years at room temperature		Accelerated life test as per protocol #zxasxz			
Complaint File Analysis— Engineering Study #987	Reliability requirement; no reactivity with drugs	Tubing cannot react with IV drug in saline solution.	Material must be PVC #avf as specified in BOM #777	Material evaluation in toxicology lab qualified PVC #avf.	Extrusion process IQ protocol #123; OQ protocol #243, and PQ protocol #544.	Supplier has been certified. Each raw material lot will be tested by independent lab and COC[a] will be issued.	Preclinical evaluation protocol #467

[a]COC = certificate of conformance.

106

■ Which elements of the original DADP will be revised? For example, if there is change to raw materials, then a possible set of DADP elements to update are:

❑ Design specifications

❑ Reliability plan and reliability testing

❑ Biocompatibility testing

❑ Risk analysis

  — SFMEA

❑ Design verification and validation

❑ IQ/OQ/PQ

❑ FDA submission

  — The regulatory affairs group shall always ask the question of whether or not the submission has been altered.

Companies should have a standard format or template that specifically addresses changes. There should be a predefined roster of functions responsible for pre- and post-approving a design change plan. From our experience in the medical device industry, a typical FDA concern is verification and validation. Firms should address both in all design change documentation, especially if design verification will be done and not design validation. A sound scientific/engineering rationale shall be stated. The same applies to process validation.

Further Reading

Gryna, Frank M., and Joseph M. Juran. 1993. *Quality Planning and Analysis*. Boston: McGraw Hill.

CHAPTER
FIVE

Measuring Design Control Program Effectiveness

This chapter brings the topic of metrics and effectiveness to the discussions about design control. As part of a business management system, internal company programs are supposed to render a return on investment. The beginning of this book through Chapter 4 discussed the economic benefits that design control and process validation can bring to the business. The claim was made that the regulation and guidelines constitute a synergistic force that, combined with other company programs, may improve the business.

The question at hand is, then, how do we measure the effectiveness of the design control program? We should remember that safety and effectiveness will lead to customer satisfaction and compliance. If we add reliability, we are then talking about premium pricing due to product differentiation. Medical device companies looking for double-digit growth say that they can differentiate from the others by bringing innovation to a given medical field. Product innovation produces premium pricing, but comes at a premium cost. Quality, reliability, and compliance with the design control requirements are not free, but when effective design control programs and systems are in place, attaining such product attributes at a reasonable cost may prove to be plausible, assuming that market analysis and industry structure strategies have been correctly defined.

Poor quality and reliability usually end up costing much more than good quality and reliability, especially in the medical device field, because doctors and other health-care givers are exposed to litigation on every procedure that they perform. This underscores the fact that one of the most important design control requirements is design input. The effectiveness of design input activity is a clear indicator of the effectiveness of the design control program. In the years to come, the ideal R&D or new-product development quality and reliability engineer should be capable of facing the challenge of working with marketing and design engineers in acquiring knowledge from customers for eventual definition of design inputs.

Design Control Program Metrics

Some of the metrics and ideas discussed in this section are more tangible than others. Any business can define design control effectiveness by one or more of the following metrics:

- The level of understanding of the customer's needs and wants

- The level of understanding of other ancillary equipment and/or drugs used with the medical device

- The accuracy of the device's intended use definition

- The device's actual ability to meet its intended use

- The complaint level predictability and its correlation to field results

- The performance level of the manufacturing process (e.g., scrap rates, rework, retest, sorting, yield, process capability, manufacturability, cost of quality)

- The number of iterations and/or prototypes in product design

- The number of redesigns

- The time to market

- The number of complaints and MDRs after launching the device[1]

- The percentage of new devices (first to market)[2]

- The percent above or below the cost of goods sold (COGS) target

- The number of open issues when the product is first released

- The status in meeting standards and goals of design quality and reliability when the product is first released

---

1. Industry management must do a better job of keeping the R&D teams in a closed loop. The typical company does not make the R&D team responsible for the consequences of their design, which leaves the team in an open loop, and they do not learn from their mistakes. If all complaints and MDRs are assessed and handled by a functional group, the R&D team will never have the chance to internalize the concepts of safety and effectiveness once the product is in the field.

2. "First to market" here means breakthrough technologies. Additionally, a company with a strategy based on innovation will need a very effective design control program because new technologies and devices must show increased performance, impeccable reliability, and economic benefits to have a chance to survive within the medical community.

As a quality system, design control effectiveness can be defined as the ability to meet the requirements stated in 21 CFR Part 820. The measurement mechanism is based on QSIT guidelines.

The following sections discuss some of the most intangible metrics[3] from the preceding list. Those not discussed here can easily be seen as tangibles with one or more ways of tracking and evaluating them.

Understanding the Customer's Needs and Wants

Customer focus is the fundamental principle behind an effective gathering of inputs for design. A typical question at this early stage is: Who are the customers? A simple answer would be all those impacted by the medical device. As an example, consider a situation in which a device manufacturer designs an excellent surgical instrument, but there are limitations with the kind of ancillary equipment needed to operate this device. As a result, the support staff and the hospital administration decide not to buy the instrument. In this example, what is in question is not the safety and effectiveness of the specific device, but a real understanding of the "use environment" and the logistics of the surgical procedure. The idea is that the total system (e.g., patient, doctor, nurse, operating room, other equipment, device, and so on) must be assessed. In this case, we see two elements that a good design control program can consider as goals that not only help with regulation compliance, but also contribute to the top and bottom lines of the firm:

- The improvements in understanding by all the people who in one way or another will be involved with the device (e.g., the patient, surgeon, doctor, nurse, clinician, installer, and so on)

- The improvements in the level of understanding that company staff personnel need to do business (e.g., how well marketing and sales know the device and its clinical effects)

Other Levels of Understanding

The level of understanding of other ancillary equipment and/or drugs to be used with the medical device is not an issue when the medical device is sold with all the accessories and components made by the same manufacturer, as when an IVD manufacturer sells its own calibrators and controls, reaction cells, and so on. The big issue comes when the medical device will interface with other equipment or drugs not under the control of the manufacturer. Care must be taken, thus, when stating the indications for use. This is another metric that can draw its performance data from field quality.

---

3. From our point of view.

Intended Use

The ability of a device to meet its intended use can be measured by evaluating the validity of the design inputs. The validity of design inputs is challenged when performing the design validation. However, field performance data tracking (e.g., complaints and MDRs) can provide better evidence and/or complementary information to design validation.

Potential Complaints

As part of design transfer, the product development team should review the hazard and/or risk analysis and the FMEAs with the firm's field quality personnel. One of the most important outputs are the potential complaints and how to classify them into categories for data analysis and corrective action. Another output is an estimation of the rate of complaints in order to establish action limits. For example, for IVD, a 1 percent rate for controls out of range during instrument calibration may be typically acceptable, whereas a 1 percent rate for broken glass bottles of the same controls may not be acceptable. Another example is the possibility that few or none of the predefined complaint categories are observed in the field, although a large number of "other categories" are received.

Another consideration is the number of complaints that are answered as "Cannot duplicate problem," "No defects found," or "Could not verify complaint." Although the possibility exists that the user or patient did not follow the procedures or the indications, a large number of these "unverifiable" complaints may be an indicator of poor understanding of the intended use of the device and its use environment. Thus, we also see a connection among many of the metrics defined in this chapter.

Summary

This chapter was intended to bring awareness to the fact that appropriate design control policy, procedures, and implementation can be measured with parameters that can also monitor the effectiveness of the business activities and operations as well as the financial performance of the firm.

PART TWO

Medical Device Reliability

CHAPTER
SIX

Medical Device Reliability Overview

The classic definition of reliability is the probability that a product will perform its intended function under specific environmental conditions for a specified period of time. The field of reliability gained major importance after World War I with impetus from the aircraft industry. During the 1940s, Robert Lusser introduced the basic definition of reliability and the formula for the reliability of a series system (Lusser 1958). The 1950s saw an increase in the use of such terms as failure rate, life expectancy, design adequacy, and success prediction. But not until the 1960s were new reliability techniques for components as well as systems developed at a faster rate. In 1961, H. A. Watson of the Bell Telephone Laboratories introduced the concept of fault tree analysis (FTA) (AMC Safety Digest 1971). Due to nuclear power reactor safety considerations, much emphasis was placed on FTA during the 1970s. Software reliability assessment has been of great interest since the mid 1970s. Much of the work done in the early 1980s concerned network reliability through the use of graphs. MIL-HDBK-217F (1991) and Bellcore (1990) are the most well known standards used for electronic equipment and system reliability prediction. These standards are used mostly during the design phase to evaluate reliability assuming random failures. In the last 15 years of the twentieth century, Markov and Monte Carlo simulation models as well as their applications in reliability and availability calculations have been considered extensively (Rice and Gopalaswamy 1993). Advances in technology have resulted in better manufacturing processes, production control, product design, and so on, thereby enabling engineers to design, manufacture, and build components and systems that are highly reliable.

FDA Classifications

The FDA classifies medical devices in three regulatory classes, depending on the level of control deemed necessary to ensure that the devices are safe and effective (FDA

1996). Several factors are considered when placing a medical device in a certain classification:

- The persons for whom the use of the device is represented or intended

- The conditions of use for the device, including the conditions of use prescribed, recommended, or suggested in the labeling or advertising of the device, as well as other intended conditions of use

- The probable health benefit from use of the device, weighed against any probable injury or illness from use of the device

- The reliability of the device

Reliability Engineering

In spite of the fact that the reliability of a device is a factor in determining device classification, and although reliability engineering has been around for a few decades now, this field of engineering is much more advanced in other industries compared to the medical device industry. Considering the fact that a medical device is a safety-critical product that must perform reliably, it is only logical and natural for customers and regulatory bodies to expect that sound reliability engineering practices have been applied by the manufacturer of that device during its design, development, and manufacture.

As indicated in Chapter 1, medical devices are used for different clinical applications under different environmental conditions with life expectancies ranging from single patient use to multiple patient use. Some devices, such as pacemakers, are implantable, whereas others, such as skin tape, may be surface-contacting. Other types of devices include those that indirectly contact the blood path; communicate directly with tissue, bone, or dentin; or directly contact circulating blood (e.g., skin staples and balloon catheters). Moreover, these devices can range from a simple mechanical technology to complex technologies, such as those for software-based microdevices or medical devices that deliver pharmaceutical agents. Due to all these reasons, it is important to focus on designing-in reliability rather than testing to prove if the product meets its design intent.

Why, then, is the medical device industry lagging behind other industries in reliability engineering? One reason is that the medical device industry is fairly young compared to other more mature industries such as automotive and aerospace. Another reason is that most of the device companies are small with few products on the market. Also, in the medical device industry, "speed to market" is a key product development metric, irrespective of whether the product is simple or complex, because competition is severe and the product life cycle typically is only about two to four years. This has led to certain behaviors in the new product development process and how the new product development organization is structured.

Problem Situations

Typical problem situations that impact device reliability include the following:

- Design engineers too often create prototypes in what is essentially a vacuum.

- Drawings and specifications are developed from the prototype, and the design is simply passed on to the production department with little concern for the product's manufacturability.

- The product development group designs and develops manufacturing processes with little concern for the long-term manufacturability of the product.

- The product development group assumes that a simple scale-up of prototype processes is all that is needed for regular production.

These approaches will almost certainly lead to the product not being manufactured consistently and reliably. Prior to passage of the Current Good Manufacturing Practice (cGMP) Quality System Regulation (21 CFR 820.30) in 1996, design failures of medical devices were estimated to have been responsible for as many as 60 deaths each year and 44 percent of device recalls. These design-related defects involved noncritical devices (e.g., patient chair lifts, in vitro diagnostics, and administration sets) as well as critical devices (e.g., pacemakers and ventilators). Also in 1990, the inspector general of the Department of Health and Human Services conducted a study entitled "FDA Medical Device Regulation from Premarket Review to Recall," which reached similar conclusions. With respect to software used to operate medical devices, the data were even more striking (Dept. of HHS 1991). A subsequent study of software-related recalls for fiscal year (FY) 1983 through FY 1991 indicated that more than 90 percent of all software-related device failures were due to design-related errors, generally, the failure to validate software prior to routine production.

Reliability Mandates

A quick search for regulations, standards, or guidance documents for medical devices that mention reliability resulted in Table 6.1. From the sample list given, the importance given to reliability by these documents is quite evident.

TABLE 6.1 Example of Regulations, Standards, and Guidance Documents That Mention Medical Device Reliability

Regulations/Standards	Description
21 CFR Parts 808, 812, and 820	Medical Devices; cGMP Final Rule; Quality System Regulation
21 CFR Parts 803 and 804	Medical Device Reporting: Manufacturer Reporting, Importer Reporting, User Facility Reporting, Distributor Reporting
FDA Draft Guidance document	General Principles of Software Validation (FDA CDRH 1997a)
21 CFR Part 820.30 (also ISO 13485)	Design Control Guidance for Medical Device Manufacturers
Medical Device Directive 93/42/EEC. Article 23, Annex I	Essential requirements for medical devices
ANSI/AIAA R-013 Reliability	Software Reliability (ANSI 1992)
ANSI/ANS-10.4 Validation	Guidelines for the Verification and Validation of Scientific and Engineering Computer Programs for the Nuclear Industry (ANSI 1987)
IEC 601-1-4 Programmable Electrical Medical Systems	Medical Electrical Equipment (IEC 2000)
IEEE 730	Software Quality Assurance Plans (IEEE 1989)
IEEE 828	Software Configuration Management Plans (IEEE 1990)
IEEE 830 Specifications	Recommended Practice for Software Requirements Specifications (IEEE 1993)
IEEE 1008	Software Unit Testing (IEEE 1987b)
IEEE 1012	Software Verification and Validation Plans (IEEE 1986)
IEEE 1042	Guide to Software Configuration Management (IEEE 1987a)
IEEE 1228	Software Safety Plans (IEEE 1994a)
ISO 9000-3	Quality Management and Quality Assurance Standards (ISO 1997)
ISO 12119	Software Packages—Quality Requirements and Testing (ISO 1994b)
UL 1998 Safety Related Standard	Standard for Software in Programmable Components (Underwriter Laboratories 1994)
FDA Guidance document	Guidance for the Content of Premarket Submissions for Software Contained in Medical Devices (FDA CDRH 1998)
FDA Guidance for industry and FDA reviewers	Medical Devices: Draft Guidance on Evidence Models for the Least Burdensome Means to Market: Availability (FDA CDRH 1999b)
21 CFR Part 864	Medical Devices; Classification/Reclassification of Immunohistochemistry Reagents and Kits
21 CFR Part 814	Premarket Approval of Medical Devices

Myths Concerning Reliability

During our interaction with many medical device professionals over the years, we have posed many questions similar to the following:

- As a quality engineer, how often have you heard design engineers say there is no need to test the product to failure?

- As a design engineer, how often have you heard from your management that once you have figured out an acceptable design, it is time for manufacturing to do the rest and make the product reliable?

- As a design control auditor, how often have you seen application, design, and process FMEA rolled into one document?

- If you are in manufacturing, how often have you felt that, if only the product development team consulted with you before cutting the tool for a plastic injection molded part, you would not have to fight fires now?

- As a quality control manager, how often have you wondered how certain specifications were established up-front?

The responses in many cases were, "Very often." Therefore, from a reliability perspective, we believe strongly that the device industry might be more concerned about the following myths:

- *Myth #1:* Using an up-front reliability engineering approach would add significant time and cost to medical device design and development.

- *Myth #2:* The use of reliability engineering tools required by regulatory agencies is sufficient to result in reliable products.

- *Myth #3:* It is easier to test the medical device at the end of the design phase if it is simple than it is to develop a plan to grow the reliability as the design progresses.

- *Myth #4:* Quality engineering as a discipline and quality engineers as a staff are responsible for downstream process control and sampling techniques and not for contributing to ensure reliability up-front in the product development process.

The reality, however, is that when reliability engineering tools and techniques are properly used, the results are favorable and cost effective. An engineering approach concurrent to product development not only will result in shortened product development time to market and reduced total life cycle cost but also will ensure product quality and reliability. Reliability engineering tools such as risk analysis, which are required by regulatory agencies, certainly help in improving the reliability of the prod-

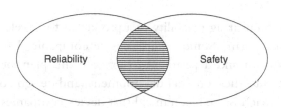

FIGURE 6.1 Relationship between reliability and safety.

uct, but often are insufficient to meet all customer needs. Testing a medical device to prove reliability without any up-front effort might prove expensive because the product may not meet regulatory and/or customer requirements. Designed-in reliability can be accomplished most effectively by integrating reliability engineering activities with other design engineering tasks throughout all phases of product development. The consequences of poor reliability efforts—customer complaints, serious injury MDR, degraded equipment performance, excessive repair costs, recalls, patient safety hazards, and even patient fatalities—are well documented. One of the primary reasons the FDA initiated design control in the new rule was to diffuse design responsibility throughout the organization and ensure shared decision making.

Reliability Versus Safety

What, then, *is* reliability, and *whose* business is it? Because reliability is quality that continues over time, it is everybody's business. Reliability has an aspect of engineering uncertainty, and hence it can be characterized by failure distributions. Thus, reliability engineering science is an integral part of product development. As mentioned in Chapter 2, reliability is a measure of the quality of the product or process design. Reliability is not necessarily the same as safety.

Figure 6.1 represents the relationship between reliability and safety. Although the safety and reliability regions of a medical device may overlap, they are not similar. Note that a reliable medical device can pose a safety risk due to other factors. The following two examples illustrate such risks:

- A diagnostic assay detecting viral antigens may be *reliable* in meeting specified ranges for disease indication but may not be *safe* because it may pick up false positives sometimes due to nonspecific protein bindings.

- A patient with an implanted *reliable* pacemaker could face a *safety* hazard when close to a cellular phone.

Summary

A simple approach to ensuring reliability of a product is to develop a disciplined approach or a "reliability plan." Some medical device companies have established a focused reliability program and department to execute this plan, but very few of the others have rolled it into their product development (and design control) process. No matter what the approach is, we recommend that device companies establish a plan to ensure reliability of their products.

The next chapter introduces the concept of the reliability plan and how it can help even small medical device manufacturers that do not have dedicated reliability resources to ensure product reliability.

Further Reading

IIT Research Institute/Reliability Analysis Center. 1991. *Fault Tree Analysis Applications Guide.* Rome, NY.

CHAPTER SEVEN

Reliability Plan

Before discussing how to create and execute a reliability plan, this chapter highlights the consequences of medical product failure to emphasize the importance of reliability. These consequences can be any or all of the following:

- Patient or user injury or death

- Delay in the clinical procedure, leading to associated costs

- Misuse of sales associate's time and energy when he or she has to explain and help the customer understand the product's performance

- Loss of product availability (or longer downtime) and cost of repair

- Increased effort to improve the design, which costs time and resources that a company could have spent on the development of new products, processes, technology, and so forth

- Regulatory intervention (483s, recalls, warning letters, and so on)

- Harm to a company's reputation that can result in reduced customer loyalty and/or market share as well as increased litigation costs and other related costs.

Consider this statistic gathered from the FDA records: 186 medical device recalls occurred in 1999 due to product malfunction or product defect (FDA 1999). Consider also the following possible situations:

- A product that was designed about three years ago is having problems in the field. Customers are complaining much more about the performance of this product. You have been looking through the test results created when the product was released initially, but could not find out how the results proved that the product met customer requirements.

■ You are the quality engineer on a new product development team and you have been asked to create a reliability plan, but you do not know where to start.

■ Your company has recently acquired another medical device company and wants to integrate the acquired company's product as soon as possible. Only limited data are available on the reliability of the product's performance in the field.

These situations, in our opinion, could have been addressed through proper execution of a product development program. A good product development program will require design and development plans (which include a reliability plan) to be created and executed to ensure product quality and reliability, even when it is not required by regulatory or certification agencies. Creating a reliability plan is not difficult. This chapter outlines the practical steps that can be taken to create a reliability plan.

Creating a Reliability Plan

Figure 7.1 indicates the primary elements of a recommended reliability plan and their links to design control elements and typical product development phases. Note that the indicated reliability focus areas are typical. Reliability activities within these focus areas are indicated for each product development phase in Figure 7.2. Each one of these reliability focus areas is explained either in this chapter or in other chapters of this book, which are appropriately cross-referenced in this chapter.

Proof-of-Concept or Feasibility Phase

Proof-of-concept or feasibility phase is usually the first phase in a product's design and development cycle. It is also the first phase where the topic of reliability must be introduced to the design team.

Product Function/Feature/Procedure Definition

The most important element in the reliability plan (as well as in the product development plan) is the definition of the quantitative and qualitative performance criteria based on customer wants and needs. This definition must include the product's function, key features, and anticipated use environment (e.g., clinical procedure). We highly recommend that adequate time and resources be spent in capturing this information. Human factor considerations must be captured to minimize risk and improve reliability.

Simple examples of sufficient and insufficient product function, feature, and procedure definitions are given in Table 7.1. Whenever possible, use quantitative definitions because they are more definitive and enable easier validation of the criteria.

Preliminary Risk Analysis and Human Factors Consideration

Chapter 8 is dedicated to the topic of preliminary risk analysis exclusively because this is a critical area of emphasis from a reliability perspective.

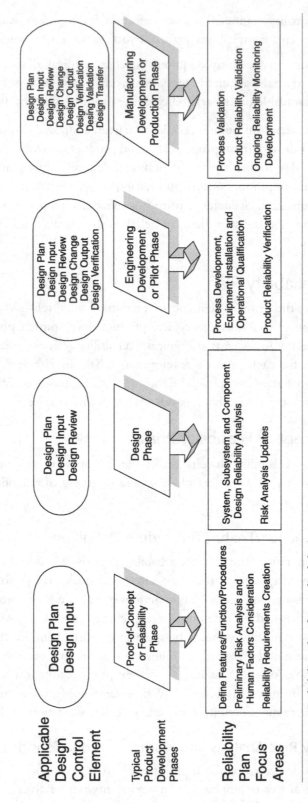

FIGURE 7.1 Reliability plan for medical devices.

Applicable Design Control Element

Design Plan Design Input	Design Plan Design Input Design Review	Design Plan Design Input Design Review Design Change Design Output Design Verification	Design Plan Design Input Design Review Design Change Design Output Design Verification Design Validation Design Transfer

Typical Product Development Phases

Proof-of-Concept or Feasibility Phase	Design Phase	Engineering Development or Pilot Phase	Manufacturing Development or Production Phase

Reliability Plan Focus Areas

Define Features/Function/Procedures Preliminary Risk Analysis and Human Factors Consideration Reliability Requirements Creation	System, Subsystem and Component Design Reliability Analysis Risk Analysis Updates	Process Development, Equipment Installation and Operational Qualification Product Reliability Verification	Process Validation Product Reliability Validation Ongoing Reliability Monitoring Development

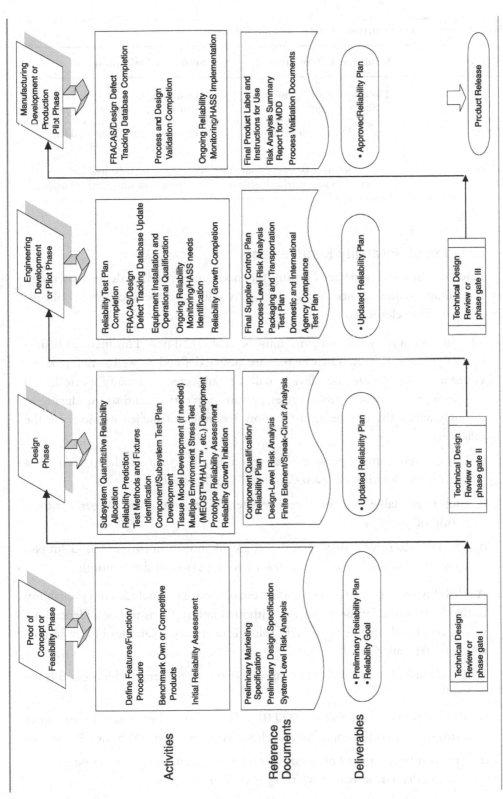

FIGURE 7.2 Recommended product reliability development plan for medical devices.

TABLE 7.1   Product Function, Feature, and Procedure Definitions

Criteria	Insufficient Definition	Sufficient Definition
Feature	The device must be visible in a dimly lit room.	The device must be visible in an operating room.
Function	The device must detect cancer.	The device must detect bladder and colon cancer using fiber optics.
Procedure	The device must be inserted into the patient to stop uterine bleeding.	The device must be inserted vaginally to stop uterine bleeding.

Creation of Reliability Requirements

Based on the medical device's feature, function, and procedure definitions, quantitative reliability requirements can be created. A reliability specification should include the following elements:

- The mission of the medical device must be established first. This includes definition of performance such that failures are clearly defined, as well as different environments to which the device will be subjected, warranty period, and definition of the mission time (cycles, pressure, force, time, and so on) during or through which the device must function. Examples of device missions are the following:

  - Contact lenses must provide clear vision and be comfortable to wear for a continuous duration of seven days with no side effects.

  - Each implantable stent must reduce arterial blockage up to 80 percent for a period of one year.

  - The HIV diagnostic assay must detect a viral antigen in patient blood samples up to 50 mol/mL at room temperature for a period of three months.

- A reliability metric or measure must be established. This includes the probability of the device completing its mission without a failure. Statistical confidence limits can be placed on this probability value, if necessary. Examples of device reliability metrics are the following:

  - Probability of survival for the mission (e.g., 95 percent at 95% confidence level)

  - Mean time between failures (MTBF): "Time" may be replaced with other terms specific to the application such as cycles (e.g., 10,000 hours, 50 cycles)

  - First-year failure rate: This metric is preferable for devices that can be repaired or those that come with a warranty (e.g., 5%)

If the product includes packaging, especially sterile packaging, reliability requirements must be created for both the device and the packaging. Subsystem and component-level reliability requirements can be created by using techniques such as reliability allocation. Details of this technique can be found in any reliability engineering textbook.

Design Phase

Device reliability is usually not constant but strongly influenced by age or usage. Therefore, it is important to know that design reliability methodologies include device age or usage to ensure successful design of a specified reliability into a device. Note that quality control can never improve reliability without making design changes. Poor quality control can potentially degrade the inherent reliability of a device.

Device failure occurs when applied load (also known as applied stress) exceeds design strength. This is applicable at the system, subsystem, and component levels.

The types of *applied load* might be friction; compression; tension; current and voltage; temperature; humidity; altitude; shock and vibration; handling, storage, and transportation stresses; electrostatic discharge (ESD) events; electromagnetic interference (EMI); operator error; and software-related stresses, among other factors. The input for applied load typically comes from customer input, established industry standards, and product benchmark data.

Design strength might be tensile strength, stiffness, fatigue strength, power or current rating (resistor), voltage rating (capacitor), assay sensitivity, viscosity, or other parameters. The input for design strength typically comes from established industry standards, product benchmark data, and testing.

The design engineer's goal is to design the medical device as well as its subsystems and components such that the design strength exceeds the applied load with appropriate safety margins or safety factors. Chapter 9 presents techniques to design reliability into medical devices during the design phase of product development.

Engineering Development or Pilot Phase

The purpose of the engineering development or pilot phase is to complete the verification of the design. The first step to accomplish that is to complete process validation.

Process Development, Installation, and Operational Qualification

Chapter 3 on process validation explains process development, installation, and operational qualification in detail. This subject is important from a product reliability point of view because a reliable process is crucial to the development of a reliable product.

Product Reliability Verification

Once design reliability is analyzed, it is important to "freeze the design." We have found that medical device companies assume that the design can be changed anytime without any impact to the reliability of the product—a sort of "design slushy" (compared to "design freeze"). We agree that the realities of product life cycle will demand design changes, but these changes must be limited to changes that do not affect the fit, form, or function of the device. Once the design is finalized, tests and analyses must be conducted to verify the reliability of the system. These range from paper analyses to bench-top engineering verification tests. Typically, these tests and analyses verify that both the hardware and software elements of a medical device meet specified reliability requirements. Also assessed is the feasibility of manufacturing the device without degrading its inherent reliability.

The finalized design can be verified using Bellcore (1990) standards to see if the design meets the intended reliability goal. The system software reliability can be verified by a combination of analyses, audits, and testing. According to the FDA, software testing is one of several verification activities intended to confirm that software development output meets its input requirements (FDA CDRH 1997a). Other verification activities include walk-throughs, various static and dynamic analyses, code and document inspections, informal as well as formal (design) reviews, and other techniques.

Figure 7.2 provides an outline of various tests that can be conducted to verify reliability. Chapter 9 outlines different test methods and analysis of test data in detail.

Manufacturing Development or Production Pilot Phase

Once the process is validated, medical devices can be built or assembled so that their reliability can be validated. Design validation means ensuring that the design consistently meets customer requirements, so there are two ways to validate the reliability of the product: The first is the testing of products in a clinical setting, and the second is the testing of products under simulated conditions.

Sample Reliability Plan

Appendix 7.1 presents an example of a reliability plan. This plan encompasses the elements of the reliability plan presented in this chapter. Although the reliability plan does not follow a rigid format, it generally should follow the approach outlined for the design and development plan.

Further Reading

Abernethy, Bob. 1996. *The New Weibull Handbook.* 2d ed. North Palm Beach: Abernethy.

Appendix 7.1.

SAMPLE RELIABILITY PLAN

Scope

This sample reliability plan outlines the steps that will be carried out to ensure the reliability of the tissue-cutting device XXX. The scope of this plan is limited to new product development elements that are applicable until design transfer to manufacturing. A separate plan will be created to monitor ongoing reliability of the device.

Products Impacted

Tissue-cutting devices XXX–a1, XXX–a2, and XXX–a3

Reliability Goal

The reliability of the tissue-cutting device system XXX must be 95 percent at a 90 percent lower confidence level for a total of 600 uses under normal operating conditions. This standard is based on marketing input of 12 uses per patient and a total of 50 patients per device with tissue thickness ranging from 3 to 5 mm.

Reliability Activities

Reliability activities include four phases, examples of which are given in the following sections.

Proof-of-Concept Phase

Normal operating conditions and the design life of the product will be the same as in the design input document for this product (Reference: DI12345). Preliminary risk analysis will be performed as specified in the risk analysis procedure (Reference: Company guidelines). Risks will be assessed using the three categories (IN, ALARP, BA) identified in the company guidelines. Risks that fall in category IN (intolerable) will be addressed through product redesign. Risks that fall under ALARP (as low as reasonably practical) will be addressed through either a design FMEA or a process FMEA. Those in the BA (broadly acceptable) category can be addressed depending on the level of risk tolerance within the company.

Design Phase

Figure 7.3 is a preliminary reliability block diagram of the tissue-cutting device XXX.

The reliability of subsystems and components in those subsystems will be allocated based on the block diagram and the system reliability goal. If applied loads are not available as design inputs for any subsystem, similar products will be bench-

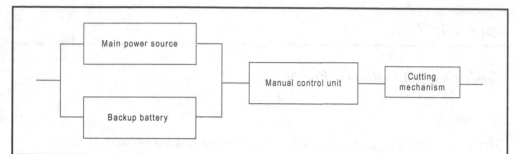

FIGURE 7.3 Reliability block diagram of the tissue-cutting device.

marked, and specifications for applied loads will be developed prior to design analysis. If applied load values are available as design inputs, safety margins and safety factors will be used (as shown in Table 7.2) to identify appropriate materials and strength properties.

All risks that fall in the BA region will be deemed acceptable risks, and no additional design or process improvements will be made. However, a safety factor of 1.5 will be used when designing for all these risks.

Finite element and tolerance stack-up analyses will be performed to ensure the dimensional fit of the components. Bellcore (1990) standards will be used to assess the reliability of the power source. Manufacturer-specified MTBF will be used as input for the back-up battery. All electronic components will be designed to run at 50 percent duty cycle.

Engineering Verification Phase

Unless specified otherwise, Weibull engineering analysis will be used for all data analyses to verify safety margins and safety factors. HALT™ (Highly Accelerated Life Testing) will be performed on the power source subsystem to detect and fix failure modes identified in the design FMEA (Hobbs 2000). Overstress testing will be performed on the total system after this activity. Packaging for the cutting mechanism and the manual control units will be subjected to overstress testing. All test plans will be developed to adequately simulate normal operating conditions. Risk analysis will be updated, and the risks that fall into category IN will be addressed through product redesign and engineering verification before moving on to the manufacturing verification phase. All failure modes will verified to be safe-failure modes.

TABLE 7.2 Safety Margins and Safety Factors for IN and ALARP Risk Categories

Risk Category and Mitigation	Safety Margin	Safety Factor (if necessary)
IN and Design FMEA	4	5
ALARP and Design FMEA	3	4
ALARP and Process FMEA	2	2

Manufacturing Verification Phase

After process validation is completed (Reference: Process validation document PV12344), a total of five systems will be assembled and tested in a preclinical setting under monitored conditions. Units will be tested to failure, and the final reliability of the product will be assessed to ensure that the product met the specified reliability goals.

CHAPTER
EIGHT

Risk Analysis and FMEA

Risk analysis is an extremely useful quality engineering methodology that builds quality up front in the design and manufacturing phases rather than having a rigorous quality inspection program. It is also a requirement for CE marking and to meet the FDA's design control guidelines. Failure modes and effects analysis (FMEA), a risk analysis tool, is one of the most powerful and practical reliability tools used in the industry to successfully improve product designs and manufacturing processes.

This chapter explains the basics of risk analysis and FMEA techniques and illustrates how to use FMEA to guide product development, design, and manufacturing. It primarily focuses on application, design, and process FMEAs. The same technique can also be used to improve the field service of a medical device.

Risk Analysis

Risk analysis of a medical device is the investigation of all available information about the product and its associated processes to identify hazards, estimate risks, and outline the steps to apply to reduce any possible risks. Figure 8.1 depicts a top-level process map for risk analysis. The risk analysis box in Figure 8.1 refers to Figure 8.2, which details the risk analysis process as defined by EN 1441 (Conformité Européen 1997). By using the inputs from the various sources indicated in Figure 8.1, the risk analysis process logically identifies potential risks posed by a medical device. Recent new standards for medical device risk analysis (EN 1441, IEC 60601-1-4 [IEC 2000], and ISO 14971-1 [ISO 1998]) as well as for FMEA (IEC 812 [IEC 1998]) specifically address the application of FMEA. Moreover, FMEA has been referenced in the FDA preproduction quality planning bulletin (FDA CDRH 1990b) and in the ISO 9001

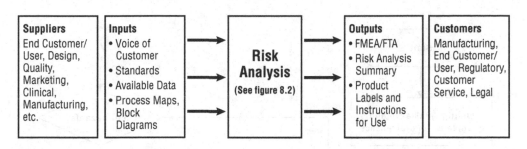

FIGURE 8.1 Process map for risk analysis.

quality standard (ISO 1994a) as an effective tool for ensuring quality product design and manufacture.

In the international standards for medical device risk analysis, risks are classified into three categories: IN (intolerable), ALARP (as low as reasonably practical), and BA (broadly acceptable). We strongly recommend that the person responsible for facilitating or completing risk analysis make sure that the output from the risk analysis captures where these risks are planned to be addressed (e.g., DFMEA [design FMEA] or PMEA [process FMEA]).

One of the steps involved in risk analysis is risk assessment. Many methods can be used to perform risk assessment, some of which are the following:

- Preliminary hazard analysis (PHA)

- Fault tree analysis (FTA)

- Software hazard analysis (SHA)

Preliminary Hazard Analysis

PHA is an inductive method of analysis whose objective is to identify hazards or hazardous situations and events that can cause harm for a given activity, facility, or system. This methodology is well suited for identifying hazards in the early development of a product when little information on design details or operating procedures is available. It can also be useful in prioritizing hazards in cases for which circumstances prevent a more extensive technique from being used. A simple "what-if" type analysis can also be used up-front for relatively simple medical devices.

Fault Tree Analysis

FTA is a top-down approach to risk analysis. It focuses on hazards (as top events) and displays potential causes. It may be very useful in complex systems in which a bottom-up approach such as FMEA would be too unwieldy. The fault tree itself is a graphic

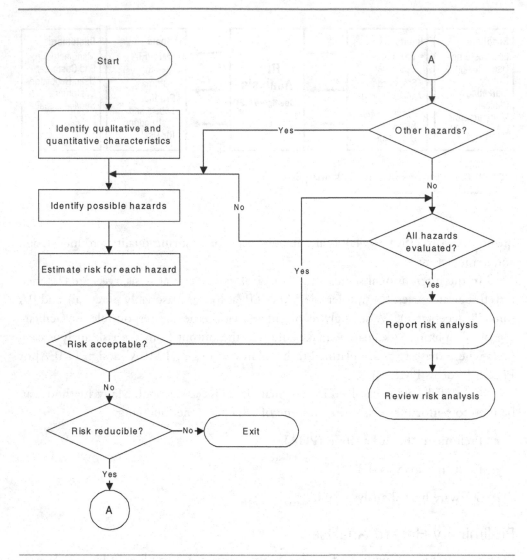

FIGURE 8.2   Risk analysis process (based on EN 1441).

model that displays the various combinations of product faults and failures that can result in the hazard. The solution of the fault tree is a list of the sets of product failures
that are sufficient to result in the event of interest. The strength of FTA as a qualitative
tool is its ability to break down a top event into basic events. This allows a safety ana-
lyst to focus preventive measures on these basic causes to reduce the probability of a
failure event. FTA can be used both to identify the underlying causes of a top event
and to calculate the likelihood of the top event if the likelihoods of the underlying
events are known.

Software Hazard Analysis

SHA is a technique for evaluating the hazards resulting from software failures. It identifies safety-critical software, classifies and estimates potential hazards, and identifies program path analysis to find hazardous combinations of internal and environmental conditions. In SHA, the software hazards are linked to:

- the software design sections where preventative measures are incorporated to eliminate the potential failure mode,

- the specific section of the software code where this hazard mitigation is implemented, and

- the software verification and validation (V&V) activities where this mitigation has been tested and found to be effective.

Failure Modes and Effect Analysis

We focus on FMEA in detail here because it is the predominant method used for risk analysis. FMEA is an easy-to-use methodology for speculating about and analyzing the effects of potential failure events on a system's design at any level of abstraction.

FMEA is an engineering analysis technique used to define, identify, and eliminate known and/or potential failures, problems, or errors from the system, design, process, and/or service before they reach the customer. The Society of Automotive Engineers defines FMEA as "a structured, qualitative analysis of a system, subsystem, or function to identify potential system failure modes, their causes, and the effects on the system operation associated with the failure mode's occurrence" (1994).

FMEA considers potential failure events—often at component or subsystem levels—and helps to determine hazard(s) that may result from those events. It is also used for prioritizing actions to reduce or eliminate the identified failure modes. In contrast to FTA, FMEA is a bottom-up approach to risk analysis. Many off-the-shelf software tools are available to complete FMEAs. Although it may be easier for a medical device company to implement any of these software tools, a concomitant difficulty lies in not only maintaining the tool but also keeping the electronic contents of the FMEAs current.

The purpose of FMEA is to analyze a system, design, process, or service to:

- identify known and/or potential failure modes and their causes as well as to determine the effects of each failure mode,

- assess the severity of failure effect and probability (likelihood) of occurrence,

- prioritize the potential failure modes identified according to the risk priority number (RPN) or risk regions,

- identify corrective actions that could eliminate or minimize the potential failure mode from occurring (including test methods, design analysis, and so on), and

- document the analysis process to support regulatory and customer service processes.

The Role FMEA Plays in Product/Process Design

FMEA facilitates communication among product/process design engineers, product/process development engineers, manufacturing/operations engineers, reliability and quality engineers, marketing professionals, regulatory professionals, and clinical research professionals. It enables these members to understand how the design or process "works." It also keeps critical items visible throughout the design stages and helps in the identification of tests needed to qualify the design or process. FMEA also provides the basis for evaluating the adequacy of changes in the product design, manufacturing process, materials, and so forth.

On the one hand, a well-constructed FMEA will provide the following benefits:

- *FMEA identifies reliability/safety-critical components and materials.* A major benefit of FMEA is that it will allow medical device companies to validate potential failure modes that are likely to jeopardize the customer or place the customer's safety at risk. This will result not only in cost savings but also in reduced time correcting problems.

- *FMEA provides a quantitative ranking of potential failure modes (Pareto analysis).* A Pareto analysis of failure modes will help the product or process design and development team focus on critical characteristics of the design or process. This, in turn, will also help in test planning that considers both normal and abnormal use conditions of the product or process.

- *FMEA is a method to track improvements based on corrective action.* Because FMEA is a logical starting point for linking documents for all changes or revisions made on the design or process, it provides a means to track improvements based on corrective action.

On the other hand, some common pitfalls and misapplications can be encountered when developing FMEAs:

- *FMEAs are used to replace an engineer's work.* We believe that an engineer's experience and intuition exceeds all statistical, quality, and reliability analyses when it comes to designing the product and assessing its risks. However, FMEA has the ability and is designed to catch the small percentage of potential problem that the engineer has not considered. FMEA should play a supporting role to assist engineers in making decisions, but not dictate the decisions. An engineer can dismiss FMEA data that are in conflict with the physics of the design; however, an engineer cannot dismiss data simply because he or she does not agree with the answer.

■ *FMEAs are used to evaluate all conceivable failure modes.* Investigating all of the conceivable failure modes can easily exceed or stretch available resources to unacceptable limits. Only failure modes that pose real challenges should be considered for a complete FMEA evaluation. The FMEA team should identify only the legitimate failure modes to be included in the FMEA.

■ *FMEAs are used to select the optimum design or process.* FMEAs can be used to select the optimum design or process, but this would be very expensive because it means developing multiple FMEAs for each design concept. FMEAs must be used to design and develop "satisfying" designs or processes.

■ *FMEA meetings are used to develop major parts of the design or the process.* FMEA development does not require large amounts of meeting time. It is cost effective and efficient when the team spends less time in meetings and more time in gathering, analyzing, testing, and validating the facts.

FMEA Basics

Before one starts to create an FMEA, it is important to understand certain necessary basics, such as the following:

■ Key terms and definitions used in FMEA. EN 1441 (Conformité Européen 1997) and ISO 14971-1 (ISO 1998) contain these definitions.

■ Customer wants and needs. The customer for DFMEA is the customer/user (and/or end customer/user) for PFMEA in the next operation.

■ Product function and process flow. Functional and/or reliability block diagrams can describe product function for DFMEA, and process flow charts (or process maps) can describe process flow for PFMEA.

■ Commitment to teamwork, continuous improvement, and a systematic bottom-up approach to failure analysis.

If these basics are present, then FMEA will become a simple but extremely effective tool for customer satisfaction. We start here with some basic questions before we discuss the details of FMEA.

What Is a Failure?

Because FMEA is all about failure modes and their effects, it is necessary to define "failure." *Failure* is the inability of a design or a process to perform its intended function. *Function* is the purpose of the design or process. This usually comes from an evaluation and analysis of the customer's needs, wants, or expectations. Important facts to remember about failure are the following:

- Failure is not limited to design or process weakness. Failure can also be due to errors made during product or process use.

- Failures are either known or potential. *Potential failures* are product or process failures that can happen when the product or process is used by the end customer/user. Because medical device failures can be caused by any number of factors and because society is litigious, we highly recommend that all failures and failure modes be captured as "potential failures" and "potential failure modes."

- Failure can happen in many forms, some of which are problems, errors, risks, concerns, or challenges.

A *failure mode* is the physical description of the manner in which an expected product or process function is not achieved. A more detailed description of failure modes is given later in this chapter in the section on the FMEA worksheet.

Who Should Be Part of the Risk Analysis and FMEA Team?

FMEA is a team-based approach. The development of an FMEA is a very interesting and fun-filled activity. Each FMEA is unique, and therefore forming the team is critical to the success of the FMEA. Once the FMEA is completed and updated and the product is released, the teams are disbanded.

Be sure that the FMEA team members are people who will be directly impacted by the changes in the design or process. We recommend that the following functions be represented in the FMEA team:

- Product design and development

- Process design and development

- Operations/manufacturing

- Reliability engineering/quality assurance

- Test engineering/maintainability

- Regulatory affairs, clinical research, and risk management

- Marketing

- Packaging and sterilization

- Customer and supplier

Who Is Responsible for the Completion of Risk Analysis and FMEA?

To ensure that risk analysis and subsequent FMEAs are developed and completed properly as well as on time, only one person must be made responsible. This must be a person who has authority and responsibility for the product or process being developed. This team leader or a project manager can either be appointed by management or selected by the team members. The team leader is responsible for:

- coordinating and facilitating the risk analysis and FMEA development process and ensuring that the team has necessary resources, and

- making sure that the team progresses toward the completion of risk analysis and FMEA and that the FMEA gets updated when changes are made to the design or process.

The team leader or project manager must not dominate the team process or influence the decisions. Someone other than the team leader or project manager must act as a scribe to document risk analyses and FMEAs, and update them as necessary. Once an FMEA team leader or project manager is identified and informed, he or she can follow the following steps to develop and complete the FMEA:

1. *Planning FMEA development.* This step includes selecting the project for an FMEA and the team members. The team then identifies the hierarchical (indenture) level at which the analysis is to be done and defines each item (system, subsystem, module, or component) to be analyzed. The team also spends time brainstorming to identify all intended items and actual functions.

2. *Investigating the failure modes, effects, and causes.* The FMEA team needs to investigate potential failure modes, their effects, and their causes. They should ask themselves three key questions:

- What are the many different ways in which a product or process can fail (failure mode identification)?

- What happens when a product or process fails (failure effects identification)?

- Why does the product or process fail (failure causes identification)?

In summary, the team must identify all legitimate potential failure modes; determine the effects and causes of each failure mode; and then classify failures by their effects on the system, subsystem, component, or process operation and mission. Once the team has answered all three of these questions, they can write their results on the FMEA worksheet, discussed later in this chapter.

3. *Determining severity, occurrence, and detection.* To help the team evaluate the failure modes, they need to quantify three aspects of a failure mode, its causes, and its effects. These aspects are severity, occurrence, and detection,[1] described as follows:

- Severity of the failure mode must be the same for device-level risk analysis and FMEA. It is evaluated based on the effect of the failure mode to the customer.

- Occurrence is evaluated based on *how often* a failure mode or its cause happens.

- Detection refers to the chance of catching the problem before it is sent to the customer.

Our recommendation is to use severity and occurrence for device-level risk analysis, and to use severity, occurrence, and detection for DFMEA, PFMEA, and so on. A rating scale is typically used to quantify all three. This rating scale can be numerical (1–10) or ordinal.

4. *Interpreting the FMEA.* Once the team members have assigned the ratings for severity, occurrence, and detection for all failure modes on an FMEA worksheet, they can proceed to analyze and interpret the FMEA. The two ways to analyze and interpret the FMEA are the following:

- *Risk priority number (RPN) method.* The RPN is calculated by using the formula

$$RPN = S \times O \times D$$

where

S is severity,
O is occurrence, and
D is detection.

The teams must assign these numbers after carefully analyzing each failure mode, and its effects and causes. RPN certainly does not stand for "randomly picked numbers"! Once these numbers are calculated for all legitimate failure modes, the failure mode with the highest RPN warrants the first consideration for analysis. The RPN approach can be considered to be both reactive and proactive. It is reactive because a failure mode has to have a high detection rating before it is considered. It is proactive because potential failure modes with high occurrence and severity ratings are dealt with up-front. We recommend that FMEA not be limited to the analysis and interpretation of RPN. An area chart must be constructed to analyze the data visually.

---

1. Note that for medical device–level risk analysis, the international standards use only severity and occurrence.

■ *Area chart method.* The area chart is created by using only the severity and occurrence ratings. The rationale for this is that even though a failure can be detected and not passed to the customer, a failure possibly can cause severe harm if and when it reaches the customer. To determine the priority failure modes, the occurrence rating is plotted on the y-axis and the severity rating is plotted on the x-axis of the area chart. An example of an area chart is shown in Figure 8.3. This area chart is divided into three regions: intolerable (high priority/risk), ALARP (medium priority/risk), and broadly acceptable (low priority/risk). These regions are defined in the international standards for risk analysis. Priority is placed, of course, on high-risk failure modes.

5. *The follow-through.* This is the crucial step in reaping the benefits of FMEAs. Once steps 1 through 4 are completed, the risk analysis and FMEA document will have the potential failure modes, their effects, and their causes identified and prioritized. Current controls or detection for mitigating these failure modes will also be identified. However, if the ability to apply necessary supporting quality engineering tools or the commitment to follow through is lacking within the team, little to no benefit can be expected further from FMEA. In fact, the FMEA documentation created without a proper follow-through can sometimes be detrimental if there is a lawsuit or a regulatory audit.

Supporting Tools for FMEAs

The development and analysis of FMEAs typically require the use of other supporting quality tools. Some of them are design of experiments, process stability and capability studies, and control charts.

Design of Experiments and FMEA

The design of experiments (DOE) is extremely useful in constructing or updating process and design FMEAs. Significant causes and effects of failure modes can be determined by using the DOE approach. It also helps in understanding the effectiveness of corrective and/or preventive actions that address the failure mode.

Process Stability and Capability Studies and FMEA

Another tool that is very useful when developing process FMEA is the process capability study. A process capability index (e.g., Cpk) can be used as a basis for creating an occurrence rating. The higher the process capability index, the lower the occurrence rating. Process capability studies can also be used when assessing the impact of corrective and preventive actions by comparing before and after results from capability studies.

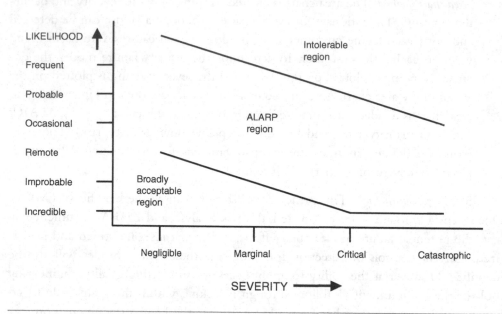

FIGURE 8.3 Area chart for risk analysis and FMEA.

Control Charting and FMEA

Control charts are applicable when performing process FMEA. Control charts help to monitor and quantify the types of failure modes and their occurrence during the construction of the FMEA. They are also useful when monitoring and assessing the impact of the corrective and/or preventive actions in the FMEA. They can identify how much the severity and occurrence of identified failure modes have been reduced. Figure 8.4 visually illustrates the recommended approach.

Risk analysis activities and their link to product development phases are outlined in Figure 8.5. The crucial activity is the updates to be performed on all risk analysis documents even after the product is released.

FMEA Templates

This section introduces templates for performing DFMEA and PFMEA and explains some columns within the templates. The documents needed for each type of FMEA are described first.

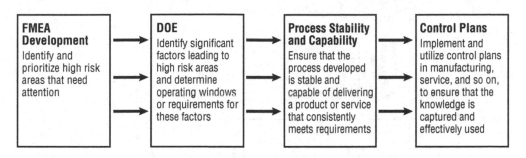

FIGURE 8.4 Recommended steps for risk analysis/FMEA follow-through.

DFMEA Template

The following components are needed to develop a DFMEA:

- Voice of the customer documentation (e.g., quality function deployment [QFD])

- Design specifications

- Reliability or functional block diagrams

- Company operating procedures for risk analysis or FMEA

- Competent personnel as identified earlier in this chapter

Fill in the design FMEA worksheet (Figure 8.6) by entering the FMEA document number, which can be used for tracking purposes, and indicating the appropriate level of analysis. Enter the name and code of the product, subsystem, or component being used.

Filling in the Worksheet

Fill in the columns as described in the following steps.

1. Item Function. Enter the name and number of the part being analyzed. Enter the function being analyzed to meet the design intent. Include information regarding the environment in which this system operates. If the item has more than one function with potential modes of failure, list all the functions separately.

2. Potential Failure Mode. Indicate the physical engineering description of the manner in which the part, subsystem, or system could fail to perform its design intent (e.g., broken gear, broken connector). The potential failure mode may also be the cause of a potential failure mode in a higher-level subsystem or system, or be the effect of one in a lower-level component. List each potential failure mode for the par-

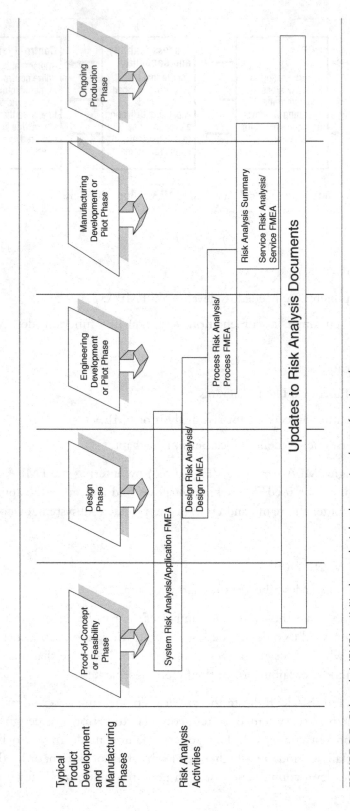

FIGURE 8.5 Risk analysis/FMEA activities by product development and manufacturing phases.

(1)	(2)	(3)	(4)	(5)	(6)	(7)	(8)	(9)	(10)	(11)	(12)	(13)	(14)	(15)	(16)
Item Function	Potential Failure Mode	Potential Effects of Failure Mode	Severity (Pre)	Potential Cause(s) of Failure	Likelihood/ Occurence (Pre)	Current Design Controls	Detection (Pre)	RPN (Pre)	Recommended Actions	Individual Responsible	Actions Taken	Severity (Post)	Likelihood/ Occurence (Post)	Detection (Post)	RPN (Post)

DESIGN FMEA

Product Name and Code:

Original Issue Date:

Document Number:

Revision (letter and Date):

Prepared by:

Responsible Engineer:

FIGURE 8.6 Design FMEA template.

ticular part and part function. A review of issues in the past and customer complaints, as well as team brainstorming are typically good starting points.

3. Potential Effects of Failure Mode. Describe briefly the effects of the failure mode on the function as perceived by the customer. Describe the effects of the failure in terms of what the customer might notice or experience (e.g., broken gear leading to "delayed clinical procedure"). State clearly if the function could impact safety or result in noncompliance with regulations. The effects should always be stated in terms of the specific system, subsystem, or component being analyzed. For example, a cracked part could cause the assembly to vibrate, resulting in an intermittent system operation. The intermittent system operation could, in turn, cause performance to degrade, and ultimately lead to customer dissatisfaction.

4. Severity. Severity indicates the degree of seriousness of the effect of the potential failure mode on the next component, subsystem, or customer if it occurs. It is directly related to the effect and it can be reduced only through a design change. Severity scales typically range from 1 to 5 or 1 to 10 (with 1 being less severe and 10 being extremely severe). The 1–10 scale seems to provide better resolution than the 1–5 scale.

5. Potential Causes or Sources of Failure. Potential causes or mechanisms of failure are indications of design weakness, the consequence of which is the failure mode. Identify and list all conceivable causes for each failure mode. List them concisely so that corrective action efforts can be aimed at pertinent causes. Some examples of typical causes are incorrect material specified, improper use, or inadequate maintenance. Some typical failure mechanisms are yield, fatigue, material instability, creep, wear, or corrosion.

6. Likelihood/Occurrence. Occurrence is the estimated number of failures that could occur for a given failure cause over the design life. The occurrence likelihood ranking has more meaning than the actual number. Removing or controlling one or more of the causes or mechanisms of the failure mode through a design change is the only way the occurrence ranking can be reduced. Similar to the severity scale, the occurrence scale also typically ranges from 1 to 10 (1 being extremely low likelihood of occurrence and 10 being extremely high likelihood of occurrence).

7. Current Design Controls. List current controls (design reviews, lab tests, mathematical studies, tolerance stackup studies, and prototype tests) being used. Three types of controls must be considered:

- Type 1: Controls that *prevent* the *cause or mechanism* from occurring

- Type 2: Controls that *detect* the *cause or mechanism* and lead to corrective actions

- Type 3: Controls that *detect* the *failure modes*

Naturally, the preferred sequence is type 1, type 2 (if type 1 is not possible), and type 3 (if both type 1 and type 2 are not possible).

8. Detection. The objective of specifying detection in a design or process is to detect a design weakness as early as possible and then compensate for the weakness. The detection scale ranges from 1 to 10 (1 being highly detectable and 10 being highly undetectable).

9. Risk Priority Number. As mentioned earlier in this chapter, the RPN is the product of the severity (S), occurrence (O), and detection (D) rankings and is a measure of design risk. The RPN is used to rank order the design concerns, and its value ranges between 1 and 1000. For higher RPNs, the team must undertake efforts to reduce this calculated risk through corrective action. In general, regardless of the resultant RPN, special attention should be given when the severity is high. RPNs by themselves have no meaning; they are used only for ranking design weakness.

10. Recommended Action. Include in this column the recommended action for each failure mode that is deemed critical. Note that an increase in design V&V actions will result in a reduction in the detection ranking only. Only by removing or controlling one or more of the causes or mechanisms of the failure through a design revision can the occurrence ranking be reduced. Finally, only a design revision can bring about a reduction in the severity ranking.

11. Individual Responsible. Identify the organization and/or individual responsible for the corrective action and the estimated completion time for the recommended action(s) (for example, Joe Engineer [Design], 2 weeks).

12. Action Taken. After an action has been implemented, enter a brief description of the actual action and the effective date. Recalculate the resulting RPN after the recommended action has been implemented to reflect the revised risk. All resulting RPNs should be reviewed, and if further action is considered necessary, repeat the process.

13.–16. Follow-up. The team leader or project manager is responsible for assuring that all recommended actions have been implemented or adequately addressed. The FMEA is a "living" document and therefore should always reflect the latest design level as well as the latest relevant actions, including those occurring after the start of production.

Process FMEA

A PFMEA is an analytical technique utilized by a manufacturing or process engineering team as a means to assure that potential failure modes and their associated causes or mechanisms have been considered and addressed. The PFMEA assumes that the product as designed will meet the design intent. Potential failures, which can occur because of a design weakness, need not be included in the PFMEA. The objectives of a PFMEA are to minimize production process failure effects on the design intent, identify and correct production problems prior to the first production run, detect process deficiencies as early as possible, and reduce the RPN.

The following components are needed to develop a PFMEA:

- Company operating procedures for risk analysis or FMEA

- Competent personnel, as identified earlier in this chapter

- Identification of needs, wants, and expectations of the customer

- Design FMEA documentation

- Process map or process flow diagram

The customer for the PFMEA should normally be seen as the end user. However, the customer can also be a subsequent or downstream manufacturing assembly operation. Sometimes it is easier to use the next operation as a customer because it might be difficult for the PFMEA team to extend the effect of a process failure mode to the end customer. We recommend the use of downstream manufacturing as the primary customer for the PFMEA and the severity table as a vehicle to link both downstream manufacturing and end user (i.e., the use of a severity table that has both effects listed side by side).

The PFMEA core team members include:

- product design and development engineer,

- process design and development engineer,

- manufacturing or operations engineer,

- quality engineer,

- line supervisor or facilitator,

- line operators, and

- facilities or maintenance specialists.

Typical questions to ask during a PFMEA development session are the following:

- Is the process or cell a critical one within the manufacturing line?

- What is the function of the process and how does the process perform its function?

- What is the expected true output parameter and effectiveness of the process?

- What constitutes a process failure?

- What raw materials, components, and subassemblies are used in the process?

- What are the process and/or assembly steps? Is there a process map available for this process?

- Is process technology proven or unproven at our company?

- Are the specified design specification requirements appropriate for the process?

- Does the process interface with other processes?

- What are the capabilities and limitations of the process operators?

- How is the process maintained, and what are the capabilities and limitations of the maintenance operators?

Filling in the Worksheet

Use the steps described in the following list to fill in the process FMEA worksheet (Figure 8.7).

1. Process Function or Steps. Provide a simple description of each process being analyzed (e.g., drilling, soldering, assembling, filling, or packaging). Also indicate the precise purpose of the process or operation. If the process involves numerous steps with different potential modes of failure, list the operations as separate processes.

2. Potential Failure Mode. The potential failure mode is the manner in which the process could potentially fail to meet process requirements and/or the design intent. It is the description of the nonconformance at that specific process step. It is assumed that the incoming parts meet their requirements.

3. Potential Failure Effect. It is preferable to describe the potential failure effects in the customer's terms, if possible. Data from customer complaints or similar FMEAs can be used to do that. If the customer is the next operation or subsequent process steps, the effects can be a nonconformance to process step requirement.

4. Severity. Severity is an assessment of the seriousness of the effect of the potential process failure mode to the customer. The discussion of PFMEA customers earlier in this chapter addresses the use of the severity assessment.

5. Potential Causes. List all possible causes that resulted in the failure mode and make sure they are described in terms of something that can be corrected or controlled. Another thing to be considered is that these causes must be specific. "Inadequate or no lighting" is preferred to merely "lighting."

Product Name and Code:									PROCESS FMEA				Document Number:			Prepared by:			
Original Issue Date:												Revision (letter and Date):			Responsible Engineer:			

(1) Process Step	(2) Potential Failure Mode	(3) Potential Effects of Failure Mode	(4) Severity (pre)	(5) Potential Cause(s) of Failure	(6) Likelihood/ Occurence (Pre)	(7) Current Process Controls	(8) Detection (Pre)	(9) RPN (Pre)	(10) Recommended Actions	(11) Individual Responsible	(12) Actions Taken	(13) Severity (Post)	(14) Likelihood/ Occurence (Post)	(15) Detection (Post)	(16) RPN (Post)

FIGURE 8.7 Process FMEA template.

151

6. Occurrence. Similar to that for DFMEA, the occurrence scale for PFMEA can rank from 1 to 10. The sources for obtaining occurrence ratings include statistical or historical data from similar processes.

7. Current Process Controls. Descriptions of the process controls that either prevent the failure mode from occurring or detect it if it occurs must be included. Examples of process controls are error proofing (poke-yoke), statistical process control (SPC), or inspections. Just as for DFMEA, three types of process controls exist for PFMEA:

- Type 1: Prevent the cause from occurring or reduce its occurrence

- Type 2: Detect the cause and lead to corrective actions

- Type 3: Detect the failure mode

8.–9. Detection. Detection ranking for a process failure mode must be assigned by assessing the capabilities of the current process controls to prevent shipment of the part having this defect. Random quality checks are unlikely to detect the existence of an isolated defect and should not influence the detection ranking. Statistical sampling is a valid detection mechanism and so is 100 percent inspection (manual and automatic). Methods for failure detection include:

- mathematical modeling,

- process verification and validation testing,

- product testing (simulated use),

- tolerance stack-up analysis,

- design of experiments,

- burn-in, and

- highly accelerated stress screening.

10.–16. Recommended Actions. If the effect(s) of a potential process failure mode could be a hazard to manufacturing personnel, corrective actions should be taken to prevent this failure mode by eliminating or controlling the causes, or appropriate operator protection should be specified. To reduce the probability of occurrence, process and/or design revisions are required. Only a design and/or process revision can bring about a reduction in the severity ranking. To increase the probability of detection, process and/or design revisions are required.

Generally, improving detection controls is costly and ineffective for quality improvements. Increasing quality control inspection frequency is not a positive corrective action and should be utilized only as a temporary measure until permanent measures are in place. In some cases, a design change to a specific part may be required to assist in the detection. Changes to the current control system may be implemented to increase this probability. Emphasis must be on preventing defects. An example would be the use of SPC and process improvement rather than random quality control checks or associated inspection.

Reviewing the Process FMEA

Review the PFMEA to ensure that function, purpose, and objective have been met. Examine the RPN and highlight the high-risk areas in terms of critical, significant, and major characteristic(s) that are important to the customer. Establish and implement a process control plan. Perform capability studies and work on processes with capability $\leq X \pm 4\sigma$ within specification.

Control Plan

A control plan is a documented comprehensive quality plan that describes the actions and reactions required to ensure the process is maintained in a state of statistical control. The PFMEA worksheet is the starting point for initiating a control plan. Control plans typically include the following elements:

- A listing of critical and significant process parameters and design characteristics

- Sample sizes and frequency of evaluation

- Method of evaluation

- Reaction and/or corrective action

Process Control Guidelines

Many steps need to be followed to implement process control. They are:

1. Identify the process to be analyzed and the opportunity for improvement.

2. Complete a process FMEA.

3. Evaluate potential and/or current measurement system.

4. Perform a short-term capability study or feasibility study.

5. Develop a control plan to identify and define critical/significant processes.

6. Train operators in control methods.

7. Implement the control plan.

8. Determine long-term capability.

9. Review the process for continual improvement.

10. Develop and implement an audit system.

11. Institute improvement program(s) such as six-sigma.

Summary

Risk analysis and management constitute a crucial step in ensuring the safety and reliability of a medical device. This chapter has provided details on risk analysis with particular focus on FMEA techniques. These techniques must be applied not only during initial product design and development but also throughout the product life cycle. The next chapter presents different methods that can be applied to design reliability into a medical device once the risks are identified.

Further Reading

Stamatis, D. H. 1995. *Failure Mode and Effect Analysis: FMEA from Theory to Execution*. Milwaukee: ASQ Quality Press.

CHAPTER NINE

Designing-In Reliability

One of the myths mentioned in Chapter 6 is that "using an up-front reliability engineering approach would add significant time and cost to medical device design and development." Chapter 7 introduced a few reliability tools and techniques that can be used during the product design phase to dispel that myth. One of the key safety and reliability techniques, risk analysis, was explained in detail in Chapter 8. This chapter introduces other, more specific techniques that can be employed during the design phase of product development to help medical device companies build reliability into their products and not lose revenue due to increased warranty claims, downstream inspection, and so on. These techniques are presented as applying to three different types of systems: mechanical, electrical, and software. The primary motivation for this presentation is that these technologies are the ones that are predominantly deployed in medical devices. However, device manufacturers that deploy other technologies such as optics or drug-device combinations can also benefit by using the principles discussed in this chapter.

Device Life Cycle

Most medical devices go through three distinct phases during their life cycle: early life, useful life, and wear-out life. Designing reliability into these products is only as effective as the ability of the design and development team to clearly understand these three phases and design products to deliver a longer useful life when the product is with the customer or user. Figure 9.1 illustrates these three phases by using what is called a "bathtub" curve. During the early-life phase of a medical device, the hazard rate decreases as the product is used. Poor design and poor manufacturing are two of the key reasons for the early higher failure rate. Companies have functional test screens and warranty periods to help reduce this failure rate. After this period, prod-

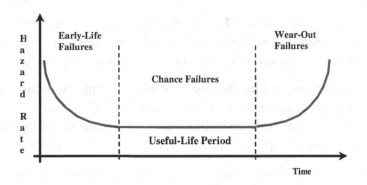

FIGURE 9.1   The bathtub curve.

ucts enter a useful-life period in which failures are purely random. If the medical device is designed for multiple-patient use and is repairable, then preventive maintenance and replacement of parts can be used to increase the useful life of a device. When the useful life period for a device ends, it will start to wear out, resulting in an increasing failure rate. Wear-out can happen due to many factors, including increased stress or load on the product, time in use, and poor maintenance.

Medical devices, like any other products, can fail even when used as intended. A look at the bathtub curve easily reveals that the useful life of a medical device can be prolonged if two approaches—failure frequency reduction and failure duration reduction—are effectively deployed. "Effective deployment" means nothing more than a committed effort for quality at all levels of the organization throughout the product life cycle.

Failure frequency reduction can be achieved through the following methods:

- Proper design of the device by selecting the right materials, right tolerances, and right system configuration (e.g., redundancy), and by making the device robust either through adequate safety margins (because higher strength means longer life) or by minimizing the part-to-part variation due to manufacturing, environment, or other factors.

- Operating the device at a lower rating (because lower stress means longer life).

- Specifying and verifying reliability requirements when components of a device are procured either off-the-shelf or through a controlled source.

- Planned and proper maintenance by having trained personnel repair or replace parts before they fail. The repair or replacement must be verified to ensure that only "equivalent" parts are used.

Failure duration reduction can be achieved through the following methods:

■ Proper design of the device by selecting the right system configuration (e.g., modular designs) so that failed modules can be removed and replaced faster. The design of personal computers is a good example of this approach.

■ Deploying adequate failure diagnostic methods, equipment, software, personnel, and spare parts to detect the location and cause of failure and to ensure that repair of parts is done faster. The failure diagnosis systems utilized by the automotive industry are examples of this.

■ Specifying and verifying availability (uptime) requirements when components of a device are procured either off-the-shelf or through a controlled source.

The best way to reduce both failure frequency and duration (and thereby improve the inherent reliability or availability of a medical device) is to build reliability into the product during the design and development phase of product development. Note that irrespective of the technology, the reliability plan with goals and risk analysis must have already been initiated at this point.

Mechanical Technology

For simple mechanical systems, load/strength interference theory is a practical engineering tool for designing and quantitatively predicting the reliability of mechanical components subjected to mechanical loading. For systems that have mechanical components, safety margins and safety factors are used to ensure reliability of the system. If *applied load* is designated as X and *design strength* is designated as Y, many design engineers will typically define safety factor as the ratio between average design strength and average applied load (Y_{Ave}/X_{Ave}) and specify the safety factor value in a range from 2 to 6. In some cases, these safety factor values have been specified to be the same irrespective of a component's function in a device. This decision, although it might appear to be insignificant compared to other activities in the design and development cycle, can actually result in two major unwanted outcomes:

■ On the one hand, the device can be *overdesigned* in that is it more than is actually necessary for safety and efficiency. This can make the device costly to manufacture because the safety factor that was chosen for some components could be very high. The end result is a lower profit margin.

■ On the other hand, the device can be *underdesigned* and expensive to maintain or replace because the safety factor chosen was very low. This could sometimes result in a recall (leading to other potential regulatory consequences), but often results in a redesign of the product.

Ideally, all the safety factors and safety margins can be assigned by using reliability allocation and some design analysis techniques such as finite element analysis. Realistically, though, we recommend that a clear link between risk analysis and safety margin/safety factor selection be established in the reliability plan (e.g., a high-risk component means a higher safety margin). For example, this can be stated in the reliability plan as, "Risk analysis including design FMEA will be used to identify critical parts and/or subsystems within a device to assign safety margins/factors. Safety margins of 4, 3, and 2 will be assigned for high-risk, medium-risk, and low-risk components, respectively." We also recommend the following order of preference of methods to be used for design analysis:

1. Safety margin calculation based on $\Phi(Y_{Ave}, \sigma_Y)/(X_{Ave}, \sigma_X)$, where Φ is the statistical distribution function (usually a normal distribution function)

2. Safety factor calculation based on $(Y_{Ave} - 3\sigma_Y)/(X_{Ave} + 3\sigma_X)$

3. Safety factor calculation based on Y_{Ave}/X_{Ave}

This approach simply enables design engineers to adequately provide margins for design strength of characteristics by taking into consideration the potential variation due to raw material, manufacturing, inspection, and so on. The end result will be extremely low interference between the applied load and design strength curves in Figure 9.2. This, in turn, means a higher reliability for the component, subsystem, or system.

For complex systems, in addition to the interference theory–based safety margin/safety factor approach, techniques such as reliability and functional block diagrams, reliability allocation, and so on can be used to analyze and improve the design at the system and subsystem levels. The FDA design control guidelines provide a com-

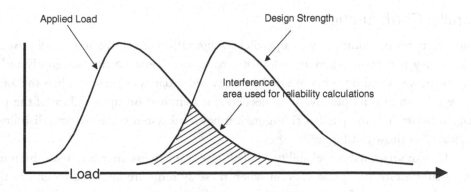

FIGURE 9.2   Load/strength interference theory.

monsense approach to product development teams to ensure that the products meet customer requirements (FDA CDRH 1996b; FDA CDRH 1996c; FDA CDRH 1997b). Because most of the mechanical components and subsystems are custom designed, design review becomes more critical to ensure that no reliability-critical characteristic is left behind. Once the production tool is cut for a mechanical part, it becomes very expensive to make changes. Therefore, it is critical that product development teams utilize techniques such as finite element analysis, injection mold flow analysis, tolerance stack-ups, metal working guidelines, and so on, and have a clear understanding of material physical, chemical, and mechanical properties of design components and subsystems. Once the components are designed, system reliability can be analyzed, depending on the configuration. This chapter presents two simple configurations and explains how reliability of a system is calculated. For more details, please refer to standard reliability engineering textbooks (for example, O'Connor 2002).

Series Configuration

Let R_1 be the reliability of subsystem 1 and R_2 be the reliability of subsystem 2. Then $P(\text{system works}) = P(\text{both subsystems work})$, where P stands for the probability of occurrence. If R_S = system reliability,

$$R_S = R_1 \times R_2$$

In general, for n subsystems in series, system reliability is given by:

$$R_S = R_1 \times R_2 \times \ldots \times R_n = \prod_{i=1} R_i$$

The important fact to note here is that the reliability of a series system is inversely proportional to the number of subsystems in the design.

Parallel Configuration

The two types of redundancy in a parallel configuration are active and standby. Active redundancy is present when all redundant items are operating simultaneously rather than being switched on when needed. Standby redundancy is present when the alternative subsystem is inoperative until needed and switched on upon failure of the primary subsystem. If the parallel configuration has n subsystems, the system reliability is calculated as shown in Figure 9.3.

Unlike series system reliability, parallel system reliability increases with the number of subsystems used. Use caution when these systems are analyzed to ensure that the final system is an optimal one and not one that is too cost-ineffective, bulky, heavy, and so on.

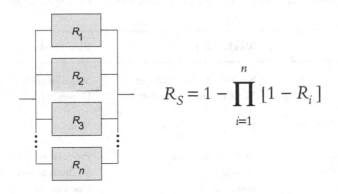

$$R_S = 1 - \prod_{i=1}^{n} [1 - R_i]$$

FIGURE 9.3 Parallel system reliability.

Electrical and Electronics Technology

Table 9.1 illustrates the differences in the application of typical attributes between electrical and electronics technology and mechanical technology as applied to medical devices. The table shows that for systems that have electronic components, a different approach is necessary to design-in reliability. Note that for electrical systems, environmental conditions (e.g., operating and storage temperatures, humidity, altitude, transportation, shock, vibration, and EMC) must be specified during the design phase because these play a larger role in determining the reliability of the electrical system. An example of the environmental requirements for an electrical system is as follows:

Operating temperature:	10 to 40°C
Storage temperature:	−40 to 70°C
Relative humidity:	90% RH at 40°C
Altitude:	0 to 10,000 feet
Mechanical vibration:	5 to 300 Hz at 2 G_{RMS}
Mechanical shock:	48-in. drop to ⅛-in. vinyl
ESD:	Up to 15,000 volts of either polarity
EMC:	The device will be used near X-ray equipment

Reliability Engineering Approaches

The following sections present some of the engineering approaches used to improve the reliability of electrical and electronic medical devices.

Reliability Improvement Through Prediction

These methods depend on failure data sources to predict reliability. Among the many different sources are the following:

TABLE 9.1   Mechanical and Electrical Technology in Medical Devices: Comparison of Typical Attributes

Attributes	Mechanical	Electrical
Product use	Single or multiple patient use	Single or multiple patient use
Sterilization concerns during design	High, due to patient contact	Depends, but typically lower
Influence of environmental conditions on design	Less	More
Failure consequence	Product discarded	Product repaired
Manufacturing volume	Very high to high	Medium to low
Reliability specification	Based on use	Based on time
Reliability demonstration tests	Less time to perform	More time to perform
Possibility for direct software interface	None	Yes
Supplier control	Source control through drawings	Specification control through industry standards
Modular designs	Difficult to implement	Easier to implement

- Government and industry databases

- Databases specific to the type of equipment/component: IEEE-STD-500 (1991) (Nuclear Energy), Bellcore TR-332 (1990) (Telecommunications), MIL-HDBK-217 (1991) (US Military), CNET RDF93 (1993) (French Telecom)

- Information supplied by the manufacturers of components

- Internal company databases on failure frequency, complaints, life tests, and product returns (both in-warranty and out-of-warranty)

These sources can be used not only to predict reliability of electronic systems but also to improve their design by selecting the right components. These components are typically selected based on predefined reliability and operating life requirements, in addition to other performance parameters. The two methods typically used for reliability prediction are parts count and parts stress. The parts count method is applied early in the design process. The steps in the parts count method are:

1. Count the number and type of components in the system.

2. Identify failure rates (or probabilities) for each component type.

3. For each type, multiply the number of components by the failure rate, and finally sum over all types.

4. Repeat steps 1–3 if there is a change in the count or type of components in the system.

The parts stress analysis method is used in the detailed design phase when individual part-level information and design stress data are available. This method requires the use of defined models that include electrical and mechanical stress factors, environmental factors, and duty cycles, among other parameters. Each of these factors must be known or determined so that the effects of those stresses on the failure rates of the parts can be evaluated.

For example, by using part failure rate, λ_p, the reliability (R) of part p at time t can be calculated. A parts stress model can be used to determine λ_p for a microcircuit as follows (Asenek, et al. 1997):

$$\lambda_p = (C_1\pi_T + C_2\pi_E)\pi_Q\pi_L$$

where

C_1 = Die base failure rate, based on the circuit complexity (e.g., number of gates);

C_2 = Package base failure rate, based on package type and complexity (e.g., pin count);

π_T = Temperature acceleration factor for die-related mechanisms (e.g., electromigration, time–dependent dielectric breakdown);

π_E = Environmental factor that accounts for all environmental effects (e.g., thermal cycling, vibration, humidity) except steady-state operating temperature;

π_Q = Quality factor that accounts for the level of screening and processing of the device; and

π_L = Learning factor that accounts for the maturity of the device manufacturing line.

Many difficulties can arise in applying these prediction models, for example:

- Some reliability handbooks, such as MIL-HDBK-217 (1991), use constant failure rate (exponential distribution) models; however, this assumption is not true in real life.

- The prediction models rely on reliability data at the component level; however, up-to-date data are extremely difficult to obtain because electronic component technology changes rapidly and component quality levels have improved.

- Actual failure rates of subassemblies supplied by outside manufacturers can be lower compared to the predicted value they provide. This may be due to assembly failures, components failures, component misapplication, inadequate timing analysis, stress-margin oversight, or other factors.

Reliability Improvement Through Physics of Failure

The physics of failure (POF) approach is based on the relationship between the device failure rate and the material properties of the device and its design. The device is tested to force failures (usually at accelerated conditions) to calculate failure rates at use conditions. This approach is preferred over the reliability prediction method due to the difficulties mentioned earlier. Sometimes the POF approach is used along with the reliability prediction method. This is done in two steps: In the first step, the POF approach is used for critical functions and critical components; in the second step, the reliability prediction method is used to modify failure rates to match field data.

Reliability Improvement Through Derating

Derating is nothing but limiting the use of a component under less severe conditions (typically about 50 percent of the maximum levels specified by the component manufacturer). Derating is the term for purposefully selecting a component so that the applied stress on the component is well below the maximum allowable stress. High reliability in electronic equipment is generally achieved by defining derating criteria (typically 50 to 60 percent of the data sheet maximum). For example, if the dissipated power in a certain application is $\frac{1}{5}$ W, it is preferable to select a $\frac{1}{2}$ W resistor to assure less stress on the device. Electronic design rules should contain specific derating instructions, and any proposed component rating above these should be subject to design reviews and justification. Derating guidelines should depend on the criticality of the device or circuit for the system functionality. Note that each electronic component will have a different derating parameter (e.g., for a resistor, power dissipation and operating temperature are critical parameters; for a capacitor, voltage and operating temperature are critical parameters).

Reliability Improvement Through Thermal Design

For electronic systems, thermal analysis techniques can also be applied to identify any hot spots and eliminate high temperatures, thereby improving overall heat dissipation. Temperature is one of the most influential variables on product reliability, and most failure mechanisms are accelerated by higher temperatures. Thermal analysis must begin as early as the design prototype stage. This should be followed by developing a good numerical model for heat flow calculations based on experimental results and model improvement.

Some of the failure mechanisms related to increased operating temperature are the following:

- Thermal coefficient of expansion (TCE) mismatch in the different materials of the chip and package

- TCE mismatch between printed circuit board (PCB) and components

- Creep in bonding materials

- Corrosion

- Electromigration

- Diffusion in the components

The Arrhenius equation has been used by the electrical industry to perform accelerated life testing and by the medical device industry to perform accelerated shelf life studies. This equation is written as

$$k = A \times \exp(-E_a/RT)$$

where

k is the rate coefficient,
A is a constant,
E_a is activation energy,
R is Boltzmann's constant (8.314×10^{-3} kJ/mol K), and
T is temperature (in °Kelvin).

This equation also provides the basis for thermal design in electronic parts. The maximum temperature within an electronic component depends on both the electrical load and the heat flow. Electrical loads on the outputs of the devices can be derated. Sufficient cooling air must be provided to allow heat generated within the devices to be dissipated at a reasonable temperature, or adequate provision must be made for convection cooling and radiation cooling. Therefore, typical techniques that are utilized to eliminate high temperatures are derating, addition of heat sinks, and increase of airflow.

Examples of thermal design include the following:

- The operating temperature and the junction temperature of the semiconductor devices are reduced below their maximum recommended limits at worst-case operating conditions. The device manufacturer's data sheet provides the recommended operating temperatures and junction temperature on the basis on its power dissipation requirements.

- PCBs are normally mounted vertically to allow cooling flow to be assisted by convection. The PCB and/or system layout must be organized to prevent secondary heating of delicate components by power devices.

Reliability Improvement Through Sneak Circuit Analysis

A sneak circuit is an unwanted connection in an electrical or electronic circuit (not caused by component failure) that leads to an undesirable circuit condition or that can inhibit a desired condition. Sneak circuits can be inadvertently designed into systems when:

- interfaces between distinct functions are not fully specified or understood,

- the design is complex or there are design constraints, or

- analysis, testing, and understanding of the system design are inadequate.

Sneak circuits are of five types:

- Sneak paths: An unexpected path is created

- Sneak opens: An unexpected open is created

- Sneak timing: Path exists at the incorrect time or does not exist at the correct time

- Sneak indicators: False or ambiguous conditions

- Sneak labels: False, ambiguous, or incomplete labels on controls or indicators

Sneak circuit analysis (SCA) is based on the identification of "patterns" within the circuit system that can lead to a sneak circuit or sneak open circuit. SCA software, which typically traces and presents all possible reverse current paths to the system designer, is available.

Reliability Improvement Through Worst-Case Circuit Analysis

Worst-case circuit analysis (WCCA) is an analysis technique that determines the circuit performance under a worst-case scenario by accounting for component variability. As shown in Table 9.1, unlike that for mechanical components, the performance of electrical components depends on environmental conditions (e.g., temperature, humidity, and so forth), component quality levels, electrical input variations, component drift due to aging, and other factors. These factors are input into the WCCA so that the output of the WCCA provides an assessment of the performance of the circuit under worst-case conditions.

Note that the electronic part characteristic database is very critical in performing a WCCA. This database would contain sources of variation and the associated variation factors. Part statistics are based on two types of component variation: bias and random. Whereas bias is predictable in direction for known inputs, random variation is not. In WCCA, bias effects are added algebraically and random effects are root sum squared.

The typical equations used in a WCCA are, for a worst-case minimum:

$$\text{Nominal} - \{(\text{nominal sum of negative biases}) + (\text{nominal square root of sum of squares of random effects})\}$$

and for a worst-case maximum:

Nominal + {(nominal sum of positive biases) + (nominal square root of sum of squares of random effects)}

Reliability Improvement Through Proper Component Selection

Many parameters must be considered before a design engineer can select an electronic component, including the following:

- Fitness for use

- Possibility of reviewing component part history before application

- Criticality of application

- Component reliability

- Supplier capability and assessment

- Availability of prediction databases

Because most electronics manufacturing for medical device companies is done by original equipment manufacturers (OEMs), the medical device companies that design the product must establish clear requirements. OEMs must also be involved upfront during the design and development of the product.

Reliability Improvement Through Other Means

Electromagnetic Interference/Compatibility Electromagnetic interference (EMI) can be defined as any natural or man-made electrical or electromagnetic event, conducted or radiated, resulting in unintentional and undesirable responses. Electromagnetic compatibility (EMC) can be defined as the capability of equipment or systems to be used in their intended environment within designed efficiency levels without causing or receiving degradation due to unintentional EMI.

Examples of EMC problems include the following:

- A computer interferes with FM radio reception.

- A cell phone interferes with operation of electronic hospital equipment.

Electrostatic Discharge Electrostatic discharge (ESD) can be defined as the discharge of charges that are built up through friction between two mediums due to their dissimilar dielectric properties. This charge build-up results in a small return current. High humidity reduces the resistance of most dielectrics, and thus will increase the return current.

Examples of ESD are the following:

- A person getting out of a car feels an electric shock when he or she closes the door.

- A person wearing wool feels a shock when moving in a vinyl chair.

Handling activities (unpacking, storing, pulling, production, test, or field maintenance) can cause ESD. When it occurs, electronic devices can be damaged in such a way that they are not immediately inoperative but fail later (latent failures). It is difficult to predict which surface will acquire which type of charges (positive or negative). Note that temperature also affects dielectric resistance, but to a much smaller degree than humidity does.

ESD protection must be built into the design by selecting suitable components and grounding techniques at the design and development phase. The manufacturing floor should be conductive and the manufacturing process must use special conductive chairs and carts. Manufacturing personnel must wear ESD straps and conductive clothing. Devices and PCBs must be kept in antistatic bags and conductive tote boxes during storage and transport.

Testability Analysis   For electronic systems, testability is one of the primary issues to be considered during the design stage. PCB assemblies can be tested only by probing the PCB surface, either by using in-circuit testers or manually. Therefore, PCBs must be designed for ease, economy, and coverage of testing, both functional and diagnostic. This may involve incorporating additional test points, as well designing with the capabilities of the test equipment in mind. A PCB that is difficult to test and diagnose can be damaged during repair and maintenance; in addition, the test results may not be reliable.

Lack of attention to testability considerations can lead to excessive production and maintenance costs. The design and test engineer must perform design review to ensure that good practices are followed and that the test coverage is adequate for the test methods and test equipment that will be used. Testability analysis can be performed by circuit simulation programs and test generation programs. Testability figures of merit are calculated by scoring the extent to which possible failure modes can be diagnosed. Test fixtures and test programs must be developed concurrently with the design. Initial prototypes can be tested manually. Initial production assemblies can be tested by using a hotbed tester. Production assemblies must be tested by using validated test fixtures and test software.

Design for Assembly   All other things being equal, electronic systems that are easy to assemble, manufacture, test, and maintain are likely to be more reliable than similar systems without these properties. A design engineer can do some things to improve reliability, for example:

- Avoid the necessity for adjustment wherever possible (e.g., potentiometers). Adjustable components are less reliable than fixed-value components and are subject to more drift. If adjustments are necessary, make sure they are easily accessible at the appropriate assembly level. Partition circuits so that subassemblies can be tested and diagnosed separately.

- Solder joints provide electrical connection and, in most cases, also provide mechanical strength to the component attachment onto the PCB. When the components are heavy, mechanical means (e.g., tie wraps, thermally conductive glue, or other attachment devices) must be provided in addition to soldering to secure the component. Solder joint reliability is a critical issue for instruments that are stored for long periods or for instruments that are expected to operate in severe environments of vibration, temperature cycling, or corrosive atmospheres.

Summary

To achieve reliable electronic equipment, it is necessary to pay attention to design, manufacture, service, and repair. During design, components must be selected carefully and their reliability determined to be consistent with the system reliability goals. In manufacturing, necessary process control steps must be implemented that do not damage the components. In service and maintenance, attention must be paid for ease of maintenance and ease of repair, and repair levels must be identified (component or subassembly or assembly). Also, field performance data must be collected to provide feedback to design and manufacturing to improve reliability.

Software Technology

Software is simply a set of instructions for a computer to execute. Medical electronic equipment is typically controlled by software to process data and monitor critical functions of the system. The Medical Device Directive (MDD) includes software in the definition of medical devices (Medical Device Directive 93/42/EEC). Recent guidelines published by the FDA on software validation as well as off-the-shelf software used in medical devices clearly indicate the agency's acknowledgment of the need to provide clear guidelines on software to medical device manufacturers (FDA CDRH 1999a).

We strongly recommend that for medical devices that contain software, the reliability plan must include software development and analysis. Elements such as software requirements specification, design and coding standards, and software validation plans must be incorporated along with hardware–software interface requirements.

Unlike hardware designs, the central problem in software design is that there are too many paths and too little time to thoroughly test software for all but the most trivial of systems. Software almost always contains defects at *every* stage of refinement. Standards for software quality assurance (SQA) were developed and used in military contracts during the 1970s. These SQA standards have now spread into commercial software development. SQA is a planned and systematic pattern of actions that are required to ensure quality in software. SQA is comprised of a variety of tasks associated with seven major activities, namely:

- application of technical methods,

- conduct of formal technical reviews,

- software testing,

- standards,

- change control,

- metrics, and

- documentation.

Software Safety Versus Reliability

Similar to the case for hardware systems, although safety and reliability are closely related for software, a subtle difference exists. In the field of medical devices, software safety examines the ways in which potential software failures result in conditions that can lead to an undesired patient or user mishap. For this reason, software failures should not be considered in a vacuum but should be evaluated in the context of an entire medical device system. Software reliability uses statistical analysis to determine the likelihood of failure occurrence. Occurrence of a failure does not necessarily result in a hazard or mishap.

Software reliability can be defined as the probability of failure-free operation of software, in a specified environment, for a specified period time. Software reliability is one of the software quality metrics. If software fails to perform consistently and reliably, it does not matter if any of its other quality metrics are acceptable.

Software Failure Modes

Software failure can be defined as the nonconformance of software to its requirements. Some software failure modes are absent data, incorrect data, incorrect timing, duplicate data, abnormal termination of event, omission of event, and incorrect event logic. It is very easy to understand why code generation is a prime source of errors because a typical program contains a large number of lines or statements. Compared to the case for hardware, a higher probability exists that correction of one software failure might introduce new errors that may, in turn, result in other failures.

All software failures can be traced to design or implementation problems because wear is not a consideration. Typically, "thousands of lines of code" (KLOC) is used as a measure for the size of the software. From a reliability perspective, it is better to use MTBF as a software reliability metric rather than defects/KLOC because the customer or user is more concerned with failures, not total error count. Because each error count within a program does not have the same failure rate, the total error count provides very little indication of software reliability.

A software program may contain many errors. Whereas some can go undetected for a long time, others have high failure rates. Even if these failures are removed, the impact on system reliability is negligible. Table 9.2, published by Bill J. Wood (1999) presents a good overview of various software failure patterns and mitigation mechanisms.

TABLE 9.2 Software Failure Patterns and Mitigation Mechanisms

Failure	Mitigation Mechanisms
Data/variable corruption	• Redundant copies; validity checking, controlled access • Cross-redundancy check (CRC) or check sum of storage space • Reasonableness checks on fetch
Hardware-induced problems	• Rigorous built-in self-test (BIST) at start-up • Reasonableness checks • Interleaved diagnostic software (see illegal function entry; data corruption)
Software runaway; illegal function entry	• Watchdog hardware • Lock-and-key on entry and exit • Bounds and reasonableness checking • Execution tread logging with independent checking
Memory leakage starves execution stream	• Explicit code inspection checklist and coding rules • Memory usage analysis • Instrumented code under usage stress analysis • Local memory control for safety-critical functions
Flawed control value submitted to HW	• Independent readback with reasonableness check • HW mechanism provides independent control/safe state • Safety supervisor computer must agree to value
Flawed display of information	• BIST with user review direction in user manual • Readback with independent software check • Separate display processor checks reasonableness
Overlapped illegal use of memory	• Explicit inspection checklist item • Coding rules on allocation and deallocation • Special pointer assignment rules

Source: Reprinted with permission from *Medical Device & Diagnostic Industry,* "Software Risk Management for Medical Devices (Table III: Failure patterns and mitigation mechanisms)," January 1999. Copyright © 1999 Canon Communications LLC.

Software Reliability Models

Software reliability depends on the following:

- Fault introduction (developed code, development process, size, software engineering technologies, tools, and experience)

- Fault removal (time, operational profile, and quality of repair activity)

- Environment (operational profile)

Software reliability models can be based on either calendar time or execution time. Models based on execution time show the best overall results. Software reliability models are based on the following:

- Hardware reliability

- Internal characteristics of the program

- Seeding models

- Other stochastic models

Hardware Reliability–Based Models

Hardware reliability–based models follow the same bathtub curve assumptions as for hardware failures. Similar to the useful-life period for hardware, the failure rate between errors is assumed to be constant. The debugging time between error occurrences has an exponential distribution with an error occurrence rate that is proportional to the number of remaining errors. In this model, one immediately removes each error that is discovered, thereby decreasing the total number of errors by one. Correction of one error, however, may inadvertently introduce other errors into the software.

Internal Characteristics–Based Model

The internal characteristics–based model computes a predicted number of errors that exist in the software. The basis of this model is a quantitative relationship derived as a function of software complexity measures. It relates specific design or code-oriented attributes of a program to an estimate of the initial number of errors to be expected in a given program.

Seeding Models

The seeding model can be used as a measure of the error detection power of a set of test cases. A program is randomly seeded with a number of known defects and the program is tested. The probability of finding j real errors of a total population of J

errors can be related to the probability of finding k seeded errors from all K errors embedded in the code.

Structured Programming

Structured programming is the approach recommended to design clear, well-defined software programs. It discourages the use of GOTO statements and uses subroutines instead. As an example, compare the following two programming codes for the same result:

- Unstructured Programming:
  □ If A > B GOTO X (line #) else GOTO Y (line #)
  □ X (line #); Y (line #)
- Structured Programming:
  □ If A > B then X (subroutine) else Y (subroutine)

Modular Programming

Modular programming breaks the program requirements into smaller separate program requirements called modules. These modules can be separately specified, written, and tested. Each module's specification must state how the module would interface with the other modules of the program. All inputs and outputs must be specified. Modular programming requires additional work in writing separate module specifications and test requirements. However, this is paid off by reduced time spent on program writing and debugging, and the resulting program is easier to understand and change. Modular programming enhances the ease of software maintainability.

Fault Tolerance

A program should be able to gracefully exit out of a fault condition and indicate the source. When safety is critical, it is also important that the program set up safe conditions before exiting to a safe state. Fault tolerance can also be provided through redundancy. Two different subroutines can be designed to handle safety-critical functions.

Real-Time Systems

A real-time system is one in which the software must operate at the speed demanded by the system inputs and outputs. For example, whereas a game program will run when executed and exactly how long it takes to complete the run is not critical, the system for a power generator must respond very quickly to operator input and must display the selected output power levels almost instantaneously. In real-time systems, the system clock synchronizes the processor input and output functions. Software must function with correct timing in relation to the system clock pulses. Timing errors are common causes of failure in real-time systems. These errors are difficult to detect, particularly by code inspection. Timing errors can also be caused by hardware or system interface problems. Logic analyzers can be used to detect timing errors and exactly pinpoint when and under what conditions timing errors occur.

Summary

This chapter has presented an overview of different techniques that can be used to design reliability into a medical device. Proper application of these techniques should result in a medical device design that is robust and reliable. Once the design is completed, the medical device needs to be verified. Reliability techniques exist that can be used to adequately verify if the output from the design and development phase of medical device product development meets the input. The next chapter discusses these reliability techniques.

CHAPTER TEN

Reliability and Design Verification

Chapter 2 mentioned that design verification is a confirmation that the design input requirements have been fulfilled by the design output. It also stated that before the FDA mandated design control to provide a sense of formality and structure, common sense drove some companies to adopt such concepts as "engineering pilot," "design pilot," and "engineering built" to verify design output. This chapter focuses on medical device reliability verification. Table 10.1 illustrates a typical design-reliability verification activity report. It clearly shows that the design output meets or exceeds the design input from a reliability point of view.

TABLE 10.1   Example of Reliability and Design Verification

Design Input		Design Output	
Environmental Conditions	**Intended Use—Based Engineering Specification**	**Designing-in Reliability**	**Reliability Verification**
The medical device must be capable of withstanding adverse use conditions in a trauma room (e.g., accidental pulling by the tubing)	The bond between the luer lock and tubing (IV line) shall be strong enough to withstand a maximum of p pounds of axial force. (Failure mode is the lock detaching from the tubing.) A safety factor of 2 will be used.	The raw material for the luer lock will be X and the solvent Y. Before inserting the tubing into the luer lock, the solvent will be applied and a curing of T minutes will be allowed.	The 99% reliability at 95% lower-bound confidence level value for the bond-strength is p_1 pounds. This results in a safety factor of 3 when compared to the applied stress.

We now introduce the concepts of reliability testing and reliability statistics. We also show how data from such testing can be used to establish evidence using reliability statistics that show the reliability of the device has been verified.

Reliability Testing

Medical devices and components must be tested for various reasons, among them the following:

- To establish evidence to meet regulatory requirements or contractual obligations

- To ensure that the costs of design and development are justified and the design output meets specified reliability requirements

- To verify if there are any new or unexpected device failure modes; to evaluate root causes for these failure modes and to plan and execute corrective action for all failure modes that need to be addressed

A typical reliability verification testing process has four steps:

1. Verify the presence of reliability requirements for the device, its subsystems, or components.

2. Plan reliability tests with proper sample sizes, test environment, and so on.

3. Conduct the tests and collect data.

4. Analyze the results to either demonstrate or improve reliability.

Reliability testing is of three types, depending on whether it is performed during the medical device design phase; the engineering or manufacturing verification phase; or the production, installation, or mainentence phase.

Reliability Tests Performed During the Medical Device Design Phase

The sample sizes available for testing in this phase are usually very limited. Some of these tests are the TAAF (test, analyze, and fix) and the HALT™ (Highly Accelerated Life Test).

TAAF

The TAAF tests are known as "quick & dirty" tests performed to learn about the design and to achieve reliability growth. Devices or parts of devices are tested under simulated use conditions to induce failures due to weak design or inadequate parts.

Design engineers usually team up with test laboratory engineers to run these tests. They analyze failures for root causes and take corrective action to fix any failures.

HALT

Unlike the TAAF tests, the HALT method (developed by Gregory Hobbs of Hobbs Engineering) systematically stimulates failures to rapidly uncover failure modes so that the root causes can be determined and fixes put in place (Hobbs 2000). These tests are typically performed on electronic boards or systems. This test method has gained popularity over the past decade because it saves time and cost over the TAAF method. However, very little mathematical or statistical theory has been established for this test method.

Reliability Tests Performed During the Engineering or Manufacturing Verification Phase

The sample sizes available for testing in the engineering and manufacturing verification phase are usually adequate and statistically valid. Some of these tests are attribute tests, fixed duration or number of failure tests, and accelerated life tests.

Attribute Tests

These tests classify test results according to qualitative characteristics such as pass/fail. Data analysis for these tests is usually performed using binomial or Poisson distributions. Although we recommend against attribute tests, we do realize that sometimes it is essential to get as much information as possible with whatever data one can collect. Kececioglu's *Reliability Engineering Handbook* (Kececioglu 1991) is an excellent reference for many of these tests and subsequent data analysis, such as the binomial-Pearson tests.

Fixed Duration or Number of Failures Tests

Fixed duration and number of failures tests are the most widely used reliability tests. Items are tested for a fixed time period or fixed number of failures. Consider this scenario in a medical device company: At the conclusion of the design and development stage in product development, four blood-glucose monitors were tested for 2,000 hours each (the life of the product), and no failures were found. The engineers concluded that the design is reliable and that the product is ready to be manufactured.

Such a scenario is not unusual. In fact, it is not unusual for products to be tested for a fixed time period, usually the life of the device, and conclusions and recommendations then made to move on to manufacturing. For the bond strength example in Table 10.1, this means that the tubes under test are pulled by applying a load of p pounds, and observations are made to determine whether the product passes or fails. If all the tubes tested withstand the load, then it is concluded that all tubes manufactured and sold are reliable. This attribute test (or success test) approach, in which the tubes are pulled with a predetermined load and the number of tubes that fail are counted, does not provide a true indication of the reliability of tubes. If the required reliability of the tube to withstand p pounds for load is 0.99 with 95% confidence, this type of testing requires 297 tubes be tested with no failures observed, as shown by the equation to determine sample size for attributes and zero failures:

$$n = \frac{\ln(1 - confidence)}{\ln(reliability)}$$

Note that engineers (and the medical device company) are wasting time and resources if they perform such "success" testing. Reliability engineering is best applied when the data collected are *failure* data, not *success* data. For the bond strength example in Table 10.1, this means that about 5 to 20 tubes are tested until any failure mode (including the expected ones) is observed and the strength value at the failure point is recorded. Even if the failure point occurs way beyond device life, it is more meaningful than suspending the test once the device life has been reached without failures. Fewer samples are needed for these tests than fixed duration tests, and the data can be analyzed through parameter estimation, confidence intervals, probability distribution fitting, and so on. Engineering evaluation of the failure modes and implementation of corrective action are more feasible with failure testing. One of the limitations of this type of test, however, is that it cannot be applied to devices that are designed for a long life (5 to 10 years). Accelerated life tests are recommended for that purpose.

Accelerated Life Tests

In contrast to fixed duration or number of failures tests, accelerated life tests are designed to induce failures rapidly via application of high stresses. In other words, as Figure 10.1 shows, whereas fixed duration or number of failures tests depend on the design strength curve to "move to the left" for a given applied load, the accelerated life tests depend on the applied load curve to "move to the right."

Data analysis is done by using correlation and regression of accelerated life data to "normal" operating stresses. Note that stresses (or loads) may be applied individually or in combination (e.g., Multiple Environment Overstress Testing, or MEOST). Our recommended approach is to apply multiple simultaneous loads because, in reality, the medical device usually sees more than one stress at any given time during its use.

Reliability Tests Performed During the Production, Installation, or Maintenance Phase

The sample sizes available for testing during the production, installation, or maintenance phase are usually adequate and statistically valid. Some of these tests are the sequential life test, acceptance tests, and Environmental Stress Screens (ESS).

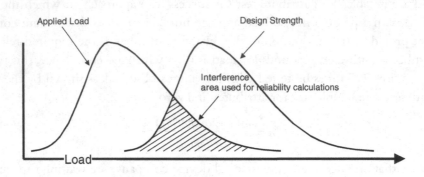

FIGURE 10.1   Load/strength interference theory.

Sequential Life Tests

In sequential life reliability testing, the devices are tested by using the process of continually assessing test results to arrive at a go/no-go decision with a minimum of testing. This test method is predominantly found in military standards and is typically used to verify reliability of production on a sampling basis.

Acceptance Tests

In acceptance tests, the medical devices are tested (usually 100 percent) to a predetermined value to evaluate conformance to design specifications before they are packaged (and sterilized, if needed). As examples, replacement joints coming off the manufacturing floor can be tested for movement, or a surgical stapler can be tested for firing staples with no load.

ESS

The ESS tests are similar to the acceptance tests with one major difference—the application of physical or environmental stresses such as physical restraints, thermal cycling, vibration, on/off cycles, and so on. This is done to eliminate defective and/or marginal parts and manufacturing defects. For example, an infusion pump can be subjected to thermal cycling and vibration to detect manufacturing defects. This test is better than simple "burn-in" tests. Highly Accelerated Stress Screens (HASS) testing is a variation of ESS testing in that it applies a combination of six degrees of freedom random-axis vibration, thermal and power cycling, among other stresses, to eliminate manufacturing defects. We recommend HASS over ESS testing when the economics are justified in terms of reduced production costs, reduced warranty costs, and so on.

Table 10.2 presents our recommended reliability tests for medical device firms. We have tried to present a realistic list of methods to select, although we do admit that to some readers the selections may come across as being conservative. When it comes

TABLE 10.2   Guidance for Reliability Test Method Selection

Primary Device Technology	Manufacturing Volume	Device Criticality (Based on FDA Classification)	Reliability Testing During Design and Development Phase	Reliability Testing During Verification Phase	Reliability Testing During Production Phase
Mechanical	Low to medium	Class I or II	TAAF	Accelerated life test	Acceptance test
		Class II or III	TAAF/Accelerated life test	Accelerated life test	Acceptance test/ESS
	Medium to high	Class I or II	TAAF	Accelerated life test	Acceptance test/ESS
		Class II or III	Accelerated life test	Accelerated life test	ESS
Electrical	Low to medium	Class I or II	HALT	Accelerated life test	Acceptance test
		Class II or III	HALT	HALT	ESS
	Medium to high	Class I or II	HALT	HALT	HASS
		Class II or III	HALT	HALT	HASS

to reliability testing, it is always "pay now or pay later." Ours is a "pay now" approach because it is cheaper to fix design problems before the medical device is released for sale or clinical trials. Note that, for a given reliability requirement, the applied stress and the sample size required for testing are inversely proportional. As such, our recommended approach can be applied by selecting fewer samples than for conventional attribute tests.

A question that always comes up when design and development teams are faced with reliability testing, no matter what the product development phase is: "How many samples should we test?" Many factors impact the selection of sample sizes, including reliability requirements, desired confidence levels (e.g., 95%, 99%), test method used, and cost of test samples. We have seen people use some general "rules" such as "assume that the test data will be normally distributed and test 30 samples" or "test 10 percent of the batch size." Based on our experience in the device industry, we provide some guidance on sample size selection in Table 10.3. Even though application of statistical methods is necessary for reliability evaluations, statistical samples are desirable but not always required during the product design and development cycle. We reemphasize that reliability testing is more meaningful when items are tested until failure and not just until their life expectancy is reached.

TABLE 10.3 Guidance for Reliability Test Sample Size Selection

Primary Device Technology	Approach	Reliability Testing During Design and Development Phase	Reliability Testing During Verification Phase	Reliability Testing During Production Phase
Mechanical	Minimum	3	5 to 20	5 to 20
	Desired	5	Statistical sample	Statistical sample
Electrical	Minimum	3	5 to 20	Statistical sample
	Desired	4	Statistical sample	Statistical sample

Reliability Statistics

We strongly believe that proper utilization of both engineering and statistics rather than just one of the two disciplines can help develop a reliable product. The classical definition of reliability is "the probability that a product will perform its intended function under specific environmental conditions for a specified period of time." Reliability statistics help the user establish this probability through test data. For simple medical devices, this is done just at the system level. For complex systems, this probability must be established all the way down to the component level.

Once the decision is made to test components, subsystems, or a complete medical device, it is highly recommended that a test protocol be created. This protocol must include various aspects of the test: the objective and purpose of the test, success criteria, necessary equipment and personnel, assumptions, sample size, data sheet, and so on. We cannot emphasize enough the need for the equipment to be calibrated and validated. Unless this is done, the test data cannot be fully understood and accepted as being indicative of the performance of the component, subsystem, or system. Note that in the example provided in Appendix 10.3 to this chapter, the sample size of 10 units might be objectionable to some statisticians as "very limited test data to fit a statistical distribution." In reality, during the design and development stages, a limited number of samples for tests is common practice. A statistical distribution can still be fit to the data to make meaningful conclusions on the reliability of the product. Many tools and techniques are available to perform a distribution fit to such "limited test data."

Once data are collected based on tests, they can be converted to meaningful information to verify if the design meets its intended engineering specifications. One way to do that is to fit appropriate statistical distributions to the data. Terms such as "cumulative failure distribution" and "reliability function" can be used with the data.

The reliability function *R(t)* can be defined as the fraction of the group surviving at time *t*. The cumulative distribution function *F(t)* is simply [1 − *R(t)*]. Typical distributions used in reliability statistics are exponential, normal, log-normal, and Weibull. Of these, the Weibull distribution is predominantly applied in reliability engineering. We focus more on the applications of these distributions rather than the statistical theory behind these distributions in this chapter.

Exponential Distribution

The exponential distribution is the most widely used failure distribution for reliability analysis of complex electronic systems. It is applicable when failure rate (λ) is constant (useful life period in the bathtub curve). The failure rate is defined as the fraction failing per unit of time (*t*) or equivalent. Examples are FIT (failures in time) rate or failures per billion hours. The simplicity of the exponential distribution is that it requires the knowledge of only one parameter for its application. The reliability of an electronic system can be calculated by using the equation:

$$R(t) = e^{-\lambda t}.$$

Engineers who assume an exponential distribution can be applied to all failure data due to its simplicity will invariably misuse this distribution, primarily because devices can fail during any one of the three phases: early life, useful life, and wear-out period, and one cannot assume that failure always occurs during a device's useful-life period.

Normal Distribution

The normal distribution is the most widely used statistical distribution in the analysis of variable data, but not in reliability engineering. The "bell curve" formed when a histogram is drawn for test data typically indicates a normal distribution fit. It is used to fit the failure times of simple electrical or mechanical systems or components such as incandescent lamps.

As an example of the application of normal distribution in reliability, suppose the failure data obtained from testing 10 bonded tubing sets is given as (in lb): 3, 2, 3.5, 4, 4.2, 4.4, 4.5, 5, 3.8, and 4.1. By using any statistical software such as MINITAB™, StatGraphics®, SAS®, WinSMITH Weibull™, or STATISTICA™ (after validating them), these data are found to fit a normal distribution at the 95% confidence level. Several conclusions can be made from these data:

- Average load to bond failure = 3.85 lb

- Standard deviation = 0.8515 lb

- Load at which 10 percent of the products would fail = 2.759 lb

- Load at which 90 percent of the products would fail = 4.749 lb

These data and the required reliability values can determine how reliable the bond strength should be.

Log-Normal Distribution

When the natural logarithms of the times–to–failure are normally distributed, then the data can be said to follow a log-normal distribution. The log-normal distribution has been found to fit cycles to failure in fatigue, material strengths, and repair data.

Weibull Distribution

Waloddi Weibull, a Swedish mathematician, developed the Weibull distribution in 1937 (Weibull 1951). This distribution is the most widely used statistical distribution in reliability engineering. Exponential and normal distributions can be modeled as special cases of the Weibull distribution, so it can even be considered as the only failure distribution that one needs to perform reliability statistical calculations. Therefore, we explain this distribution in detail here.

The advantages of the Weibull distribution include the following:

■ The ability to work with small sample sizes. Even when there are only three or four failure points in a test, Weibull analysis can be performed to obtain meaningful estimates of reliability. If there is prior history for a current test with no failures, Weibull analysis can be performed to deal with such a no-failure situation.

■ As mentioned earlier, the Weibull analysis also fits the variety of life data sets (early life, useful life, or wear-out life)

■ The Weibull parameters allow the user to provide an engineering explanation to the failure data and the failure mode that are being analyzed.

The Weibull distribution can be described using three parameters:

β (shape parameter or slope of the Weibull plot)
η (scale parameter or characteristic life, defined as the time at which 63.2 percent of the items would fail)
t_0 (location parameter or minimum life, defined as the time before which the items will not fail)

These parameters are obtained by fitting a Weibull distribution to the failure data by following the steps given below:

1. Collect the time to failure data.

2. Arrange the data in rank order (i.e., from lowest to highest value).

3. Establish median ranks for each failure (use Appendix 10.1 at the end of this chapter).

4. Plot all the points on a sheet of Weibull probability paper with the *x*-axis being the time to failure and the *y*-axis being the median ranks.

5. Draw a line through these points.

6. Calculate the slope and identify the values of characteristic life and minimum life by using the definitions given previously.

Weibull Plotting Example

We now illustrate these steps with an example. By using the data from Appendix 10.2, rank the cycles to failure and obtain the median ranks from Appendix 10.1.

In Appendix 10.2, we have manually prepared a Weibull plot by using special Weibull paper. First make sure that a best-fit straight line is drawn. Then, estimate the β parameter by drawing a straight line from the estimator point (upper left side) perpendicular to the best-fit straight line. In this case, we have estimated β to be 2.70. (In Appendix 10.3 we present a more accurate estimation by using WinSMITH Weibull™ software.) To calculate the characteristic life η, you can do one of two things:

- Use the dashed horizontal line in the plot. The estimate of η is based on finding the interception of this dashed line with the best-fit straight line. The "eye-ball" estimate is around 38 cycles.

- Another way to estimate η is to look at Table 10.4 and see that the seventh ranked value has a cumulative probability of failure of 64.4 percent. This seventh rank is equivalent to 35 cycles. Thus we can assume that the estimator is a bit less than 35 cycles.

TABLE 10.4 Weibull Plotting Example

Order	Cycles to Failure	Median Rank
1	14	6.6
2	19	16.2
3	20	25.8
4	25	35.5
5	30	45.1
6	30	54.8
7	35	64.4
8	46	74.1
9	47	83.1
10	48	93.3

Interpretations

Once the Weibull plot is completed, it can be interpreted as follows:

■ If the slope is between 1.5 and 3.0, the product tested shows an early wear-out phase.

■ If the slope is between 1.0 and 1.5, the failures are usually random.

■ If the slope is less than 1.0, then the product is in its infant mortality phase.

The reliability of the medical device or product tested is obtained in the following manner. Suppose the required reliability is 0.99 (99 percent). Draw a line from the value on the *y*-axis where the cumulative probability of failure = 0.01 (since 0.01 = 1 − 0.99) to the fitted line. Project a line from the intersection of these two lines to the *x*-axis and read the value. This value is the time at which 99 percent of the products will not fail.

Note that a two-parameter Weibull (with no minimum life or t_0) typically is used to evaluate the reliability of products. Three-parameter Weibull plots are used only when clear engineering explanations exist for why a product will not show a particular failure mode when time = 0 (e.g., fatigue failure of metals at 0 cycles). Other advanced interpretations of Weibull plots are possible (e.g., "dog-leg" in a Weibull plot, bi-Weibull, confidence levels) but require more advanced experience and knowledge in reliability and Weibull applications. Robert B. Abernethy's *New Weibull Handbook* (Abernethy 1996) is an excellent reference material on this topic.

As far as using a software program for reliability analysis, we strongly feel that the use of specialized software such as WinSMITH Weibull™ from Fulton Findings is a much better option than generic statistical software such as MINITAB™. Our reasons include the ability of the special software to perform advanced Weibull analysis and the methods used to arrive at the best distribution line that fits the data. In Appendix 10.3 we provide an example of Weibull analysis using WinSMITH Weibull™ software.

Further Reading

Dodson, Bryan. 1995. Weibull Analysis. Milwaukee: ASQ Quality Press.

Fries, Richard. 1991. *Reliability Assurances for Medical Devices, Equipment, and Software*. Buffalo Grove: Interpharm Press.

Kapur, Kailash C., and Leonard R. Lamberson. 1977. *Reliability in Engineering Design*. New York: John Wiley and Sons.

Appendix 10.1 Median Rank Values for Rank-Ordered Data

Rank Order	\multicolumn{20}{c}{Sample Size}

Rank Order	1	2	3	4	5	6	7	8	9	10	11	12	13	14	15	16	17	18	19	20
1	50.0	29.2	20.6	15.9	12.9	10.9	9.4	8.3	7.4	6.6	6.1	5.6	5.1	4.8	4.5	4.2	3.9	3.7	3.5	3.4
2		70.7	50.0	38.5	31.3	26.4	22.8	20.1	17.9	16.2	14.7	13.5	12.5	11.7	10.9	10.2	9.6	9.1	8.6	8.2
3			79.3	61.4	50.0	42.1	36.4	32.0	28.6	25.8	23.5	21.6	20.0	18.6	17.4	16.3	15.4	14.5	13.8	13.1
4				84.0	68.6	57.8	50.0	44.0	39.3	35.5	32.3	29.7	27.5	25.6	23.9	22.4	21.1	20.0	18.9	18.0
5					87.0	73.5	63.5	55.9	50.0	45.1	41.1	37.8	35.0	32.5	30.4	28.5	26.9	25.4	24.1	22.9
6						89.0	77.1	67.9	60.6	54.8	50.0	45.9	42.5	39.5	36.9	34.7	32.7	30.9	29.3	27.8
7							90.5	79.8	71.3	64.4	58.8	54.0	50.0	46.5	43.4	40.8	38.4	36.3	34.4	32.7
8								91.7	82.0	74.1	67.6	62.1	57.4	53.4	50.0	46.9	44.2	41.8	39.6	37.7
9									92.5	83.1	76.4	70.2	64.9	60.4	56.5	53.0	50.0	47.2	44.8	42.6
10										93.3	85.2	78.3	72.4	67.4	63.0	59.1	55.7	52.7	50.0	47.5
11											93.8	86.4	79.9	74.3	69.5	65.2	61.5	58.1	55.1	52.4
12												94.3	87.4	81.3	76.0	71.4	67.2	63.6	60.3	57.3
13													94.8	88.2	82.5	77.5	73.0	69.0	65.5	62.2
14														95.1	89.0	83.6	78.8	74.5	70.6	67.2
15															95.4	89.7	84.5	79.9	75.8	72.1
16																95.7	90.3	85.4	81.0	77.0
17																	96.0	90.8	86.1	81.9
18																		96.2	91.3	86.8
19																			96.4	91.7
20																				96.5

Appendix 10.2   Weibull Probability Chart

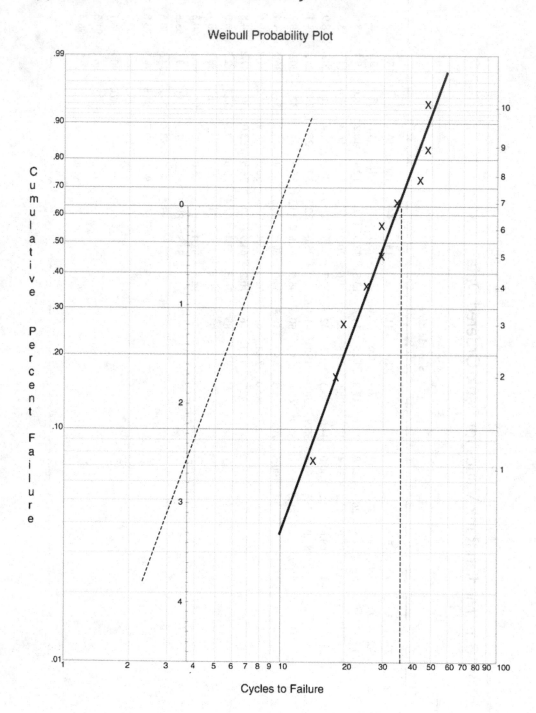

Weibull Probability Plot

Appendix 10.3   Reliability of a Mechanical Tissue-Cutting Device

The reliability requirement for a mechanical tissue-cutting device is 0.95 at a 95% confidence interval for 10 cycles under normal use conditions. To verify the reliability, 10 units were tested until failure, and the failure data shown in Table 10.5 were obtained. All failures displayed the same failure mode (pinion breakage).

Because all failure modes were the same, the engineer responsible for verifying reliability can plot one Weibull chart by using WinSMITH Weibull™ software before analyzing it. (*Note:* If there is more than one failure mode, one Weibull plot must be created for each failure mode. Data for other failure modes are "suspended" to create these plots.) The software analyzes the data and selects the appropriate distribution that fits the data (Figure 10.2). Note that the higher the r^2 value, the better the distribution fit is for the data. Because a two-parameter Weibull distribution is found to be the optimum distribution, the following values are obtained from the plot (Figure 10.3):

Slope (β) = 2.69
Characteristic life (η) = 35.34 cycles
Reliability at 10 cycles = 90 percent with 95% confidence

Based on the estimated Weibull parameters and the reliability values, the following conclusions can be reached:

- Because the reliability of the product at 10 cycles is 90 percent with 95% confidence, the device *does not* meet the reliability requirements. After 35.34 cycles, 63.2 percent of all the products would fail. The software also generates a Weibull Analysis Report as shown in Figure 10.4.

TABLE 10.5   Failure Data

Device Number	Cycles to Failure
1	19
2	14
3	30
4	20
5	25
6	35
7	30
8	48
9	47
10	46

Distribution Analysis (Regression)

Set 1 - 1

Weibull [t0 = None ... 2 Parameter]
Correlation(r)=.980816 r^2=.962 ccc^2=.8646 r^2-ccc^2= .0974 (Okay)
Characteristic Value=35.34 Weibull Slope=2.69 Method=rr

Weibull [t0 = 8.418602 ... 3 Parameter] [Scale Not As Recorded]
Correlation(r)=.9864076 r^2=.973 ccc^2=.9333 r^2-ccc^2= .0397 (Okay)
Characteristic Value=26.33 Weibull Slope=1.751 Method=rr/t0^

LogNorm [t0 = None ... 2 Parameter]
Correlation(r)=.9746794 r^2=.950 ccc^2=.8796 r^2-ccc^2= .0704 (Okay)
Log-Mean Antilog=29.1 Std. Dev. Factor=1.567 Method=rr

Normal+ [t0 = None ... 2 Parameter]
Correlation(r)=.9695359 r^2=.940 ccc^2=.8796 r^2-ccc^2= .0604 (Okay)
Mean=31.4 Std. Deviation=13.08 Method=rr

Optimum Distribution: Weibull [t0 = None ... 2 Parameter]

FIGURE 10.2 Results of data analysis by using WinSMITH Weibull™ software.

Source: from SuperSmith™ software by Fulton Findings™, contact info: WeibullNews.com
(WinSMITH™ Weibull is a component program of SuperSMITH)

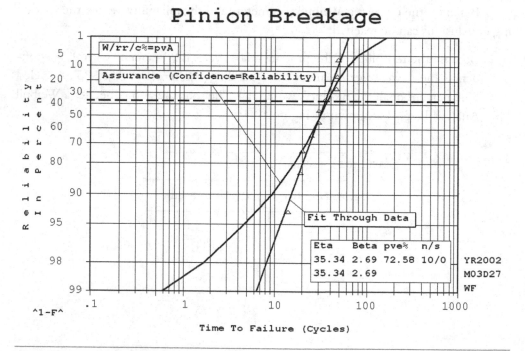

FIGURE 10.3 Weibull plot from WinSMITH Weibull™ software

Source: from SuperSmith™ software by Fulton Findings™, contact info: WeibullNews.com
(WinSMITH™ Weibull is a component program of SuperSMITH)

By WF - Set 1 - Fit Through Data   Date: M03-D27-YR2002
Pinion Breakage
Prr-Value(%)=74 r^2=.962 ccc^2=.8538 pve%=72.59 (Okay)
Characteristic Value=35.34 Wiebull Slope=2.69 Method=rr
Mean=31.43 (<31.89) Std. Deviation=12.6
Point Quantity=10 (susp=0)
c%=pvA

B%	Time (Cycles)	Assurance (Confidence=Reliability)
1	6.39	.585
2	8.284	1.701
5	11.71	4.705
10	15.31	9.48
20	20.23	16.36
50	30.84	31.12

FIGURE 10.4   Weibull Analysis Report from WinSMITH Weibull™ software

Source: from SuperSmith™ software by Fulton Findings™, contact info: WeibullNews.com
(WinSMITH™ Weibull is a component program of SuperSMITH)

- ■ The failure mode is pinion breakage, and it happens during the early wear-out stage of the product's life. The product has to be redesigned to increase the reliability so that the wear-out occurs after 99 percent of the products meet the reliability requirement of 10 cycles with 95% confidence.

CHAPTER ELEVEN

Reliability and Design Validation

The ultimate litmus test for any newly designed and developed product is its performance at the hands of customers. This is, of course, true for medical devices as well, but because the consequences of product malfunction or failure for these devices can be severe, performance is a more serious consideration. We have heard from design engineers that if the reliability of a medical device is verified on a bench-top, then there is no need for reliability testing under actual or simulated use conditions. We do agree that it is almost impossible to perform medical device *failure testing* (taking each medical device to failure) under *actual* use conditions due to its clinical significance. We also accept that it is extremely difficult to perform failure testing for capital equipment medical devices (CT scans, power generators, and so forth) under simulated conditions (but not stimulated conditions such as HALT) due to the long test time required. However, we emphasize to those engineers that it is necessary to perform *functional* testing under actual conditions (even if the device is class I or class II) before the device is released for sale, and, where possible, to perform *failure* testing under simulated conditions (e.g., in animal or tissue laboratories). Both functional and failure testing are part of reliability testing. When we analyze the definition of reliability and design validation, a strong overlap between the two is obvious.

We can examine this overlap by comparing the elements that define reliability and design validation.

- Reliability
  - Reliability is quantified in terms of probability.
  - Function or intended use must be defined.
  - Environment or operating conditions must be defined.
  - There is an operating time between failures.

■ Design validation

  ❑ Design validation shall be performed under defined operating conditions on initial production units, lots, or batches, or their equivalents.

  ❑ Design validation shall ensure that devices conform to defined user needs and intended uses and shall include testing of production units under actual or simulated conditions.

  ❑ Design validation shall include software validation and risk analysis.

From this comparison, we readily see the first two points under design validation make reference to environment or operating conditions and intended use. What is left is the probability and operating time between failures. Here is where device software (if applicable) and risk analysis come into play. When performing the preliminary hazard analysis (PHA), the design and development team has basically exposed the design or concept to the potential hazards of the clinical or medical application. Thus, depending on the severity and frequency of such hazards, redesign and/or reconsideration of design goals is performed. Frequency in risk analysis and/or FMEA basically captures "probability" and "operating time between failures," as stated in the definition of reliability. For example, a short life or small mean time to failure for the power generator of a pacemaker will be deemed unacceptable as part of the design goals (i.e., the higher likelihood of failure during the device's life).

Another example is that 35 years ago a pacemaker susceptible to stray electromagnetic forces was deemed acceptable because microwave ovens were not as popular then as they are today. Thus, we see that the "operating conditions" or environment is dynamic in nature, and thus reliability and/or reliability requirements may change with the rest of our technology. Today, modern pacemakers are shielded from stray electromagnetic forces and have a back-up mode (e.g., parallel redundancy) that takes over if a really strong electromagnetic field disrupts the main circuit's programming (Abben 1999).

Why does design validation require the use of reliability tools and not typical quality assurance tools? In addition to the explanation above, let us look at the typical QA functions when it comes to validating a device, namely, inspection and testing. Without considerations for design robustness and worst-case operating conditions, what product inspection will ensure proper functioning of a device? What about lot acceptance testing that does not overstress the device? If a random sample of products is tested to its life under controlled conditions, how reliable is the entire population of devices? Note that in the case of sampling, there is always a beta error (i.e., the probability of shipping a failing device to the market when the results of inspection and testing have indicated it was okay).

The preceding discussion points to the reality that reliability and design validation are indeed interrelated. Whereas reliability life testing typically takes the product beyond its intended life (and use), design validation testing typically expects the device to perform as intended until the end of its life. Reliability testing also provides answers to questions such as, "If a total of 100,000 devices are to be sold, how many of them would fail when used as intended?" or "What percent of devices would fail within the warranty period?" Thus, it is easy to see that reliability testing not only helps the company meet the design validation regulations, but also allows the company to plan for warranty and product service (if needed).

Now that we believe that we have convinced the readers about the importance of design validation and its interrelationship with reliability, we will answer a few important design validation–related questions that we have faced in the past:

- *How does one perform design validation–related reliability testing?* As much as possible, test the products under actual intended use conditions and collect failure data. The next best choice would be to test the devices in a simulated use condition. This can be a bench-top or animal or tissue laboratory test. Always create a written protocol with clearly defined success criteria, and use the guidance table (Table 10.2) provided in Chapter 10 to select the appropriate test approach.

- *How many samples does one need to test?* Use Table 10.3 provided in Chapter 10. Whenever possible, make sure that the sample size has sufficient statistical justification. For derivative products, use WeiBayes Weibull analysis–based substantiation testing methods provided in Abernethy's handbook (Abernethy 1996) to calculate sample sizes because Weibull parameter information based on predicate device reliability verification testing would be useful.

- *Does one need to train the operator of the device?* Because reliability testing is based on "intended use of the device," one must try to simulate the actual use condition, even if it means not training the operator of the device.

- *How does one analyze data?* Weibull plots discussed in Chapter 10 can be used to analyze the data. For devices that are repairable, it is recommended that failure rate at the end of warranty period (if applicable) and repair rates be calculated. As mentioned in Chapter 10, make sure that there is one Weibull plot for each failure mode.

- *What should one do if reliability testing for design validation failed?* If the reliability testing failed, take a close look at the data and evaluate the failure mode, the product's design life when the failure occurred, the applied stress when the failure occurred (to see if it is in the overstress region), and other contributing factors (e.g., power failure, operator abuse). Consult with clinical and reliability engineering experts to understand these factors, to make a determination if the failure mode is safe and is highly unlikely to occur, and to assess if corrective action is necessary.

■ *Is it possible to still release the product to the market if failures occur beyond the product's life?* It is possible if there is sufficient evidence as well as clinical guidance. Refer to the previous question.

Design validation or design verification without a statement about risk and reliability is as incomplete as a clinical study that lacks a statement about confidence. Reliability-based design validation testing is important and should not be avoided or stopped due to cost considerations. If the choice is between pay now (test cost) or pay later (recall, warning letters, lost sales, and so on), quite obviously, functional and failure testing under actual or simulated conditions is always preferable. This type of testing will prove that the product is reliable and that the design can be validated.

CONCLUSION

As early as 1984, the FDA had identified lack of design controls as one of the major causes of medical device recalls. Now, implementation of design controls is part of the FDA regulation, and manufacturers should expect FDA inspectors to look for evidence of implementation. That is, FDA inspectors evaluate the process utilized by medical device companies to design and develop products (device, labels, and packaging) by looking at the companies' policies and procedures. Unless it is obvious, they are not evaluating safety and effectiveness of the device under design control. Creating the procedures and instructions for design control implementation is very easy compared to the actual implementation and understanding of what can really be expected to be achieved. Throughout this book, we have attempted to explain the design control requirements, the benefits of design control to business, practical means for adoption and implementation of design control, and the interrelationship of design control to all other quality systems. Now it will be up to the medical device firms to take advantage of the set of tools provided in the second part of this book to do what companies in other industries have already been doing to create market differentiation and brand names.

The FDA design control guidelines are a great contribution to make the medical device industry more dynamic and competitive. Some medical device companies are already enjoying the benefits of applying the tools far beyond what the regulations require. But the greatest benefits will come in the years ahead. With the emergence of new treatments and new discoveries, and the application of high technology to healthcare, the design of medical devices will not be as simple as it is today. We can expect more applications involving electronics, optoelectronics, smart and adaptive drug delivery systems, and extensive use of software code, among other technological advances. With the emergence of the Internet, we can also imagine "real-time product complaints" and feedback from users to manufacturers. Those companies that have already developed effective design control systems and are living up to those standards have walked the first couple of miles of the new road to the future of healthcare. The complexities of the devices of the future can be managed only with the disciplines and tools of design control. Quality engineering disciplines such as reliability, voice of the customer analysis, design of experiments, software quality, and others will play even more of a major role to help medical device companies bring quality products to market sooner, and at the same time meet the intent of design control guidelines.

ACRONYMS AND DEFINITIONS

AFMEA Application FMEA, used to identify and minimize risks associated with failure modes during device application from the time it is picked from the shelf or installed on site.

ALARP As low as reasonably practical.

ANOVA Analysis of variance.

BIST Built-in self-test.

BOM Bill of materials.

CABG Coronary artery bypass graft.

CAF Change approval form. This is an industry term used to define, control, approve, execute, and document changes to design, processes, specifications, testing, and so on. Also known as ECN or ECO and similar names.

CAPA Corrective and preventive action.

CDRH Center for Devices and Radiological Health.

CE Conformité Européen. Certification required to market medical devices in Europe.

cGMP Current GMP.

COC Certificate of conformance.

COGS Cost of good sold.

CP Corporate Procedures or Company Procedures. Also known as SOP or OP.

Cpk Process potential capability ratio.

CPM Critical path method.

CRC Cross-redundancy check.

CSA Canadian Standards Association.

CT Computer tomography.

DADP Design and development plan.

DEHP Di-ethylhexyl-phthalate, a toxic plasticizer used to make vinyl medical products. DEHP is the primary phthalate plasticizer used to make PVC medical devices (e.g., blood bags, intravenous bags, and medical tubing) soft and flexible.

Design To devise for a specific function or an end.

Development Once there is a concept, to plan and construct the prototypes and manufacturing process.

DFMEA Design FMEA focused on the product.

DHF Design History File. A compilation of records describing the design history of a finished device. This file should show that the device was designed according to the DADP.

DHR Device History Record. A compilation of records containing the production history of a finished device. In industry, it is typically called the "batch record."

DMR Device Master Record. A compilation of records containing the procedures and specifications for a finished device. In industry, this can be seen as all "live" documents needed to make a medical device.

DOE (modeling) Design of experiments. For *modeling*, DOE refers to those experiments aimed at obtaining a prediction model based on input or independent variables that are known to have an effect on the output or dependent variable. For *screening*, DOE refers to the initial experimentation aimed at identifying the most likely factors or input variables that can affect a given output. Typical screening experimental design are Taguchi and some fractional factorials.

ECN Engineering change notice.

ECO Engineering change order.

EMC Electromagnetic compatibility.

EMI Electromagnetic interference.

EPROM Electronically programmable read-only memory.

ESD Electrostatic discharge.

ESS Environmental Stress Screens.

FDA (US) Food and Drug Administration.

FITs Failures per billion hours.

FMEA Failure modes and effects analysis.

FRACAS Failure Reporting and Corrective Action System.

FTA Fault tree analysis.

FY Fiscal year.

GHTF Global Harmonization Task Force.

GMP Good Manufacturing Practice.

GR&R Gage Repeatability and Reproducibility.

HALT Highly Accelerated Life Testing.

HASS Highly Accelerated Stress Screens.

Human factors Discipline that encompasses the various methods used to improve compatibility between human beings and the medical device.

HVAC Heating, ventilation, and air-conditioning (system).

HW Hardware.

ICT In-circuit test.

IDE Investigational Device Exemption.

IN Intolerable.

IQ Installation qualification.

IT Information Technology.

IV Intravenous.

IVD In vitro diagnostics.

KLOC Thousands of lines of code.

MA Broadly acceptable.

Manufacturability The ability of a designed manufacturing process to produce an output.

MCA Measurement capability analysis.

MDD Medical Device Directive.

MDR Medical device report.

Medical device According to section 201(h) of the Food, Drug, and Cosmetics Act. A, "an instrument, apparatus, implement, machine, contrivance, implant, in vitro reagent, or other similar or related article, including a component, part, or accessory, which is recognized in the official National Formulary, or the United States (U.S.) Pharmacopeia, or any supplement to them, intended for use in diagnosis of disease or other conditions, or in the cure, mitigation, treatment, or prevention of disease, in man or other animals, or intended to affect the structure or any function of the body of man or other animals, and which does not achieve any of its primary intended purposes through chemical action within or on the body of man or other animals and which is not dependent upon being metabolized for the achievement of any of its primary intended purposes."

MEOST Multiple Environment Overstress Testing.

MPS Master production schedule.

MRI Magnetic resonance imaging.

MTBF Mean time between failures.

MTTF Mean time to failure.

MTTM Mean time to maintenance.

MTTR Mean time to repair.

NCCLS National Committee for Clinical Laboratory Standardization.

OEM Original equipment manufacturer.

OP Operating Procedures.

OQ Operational qualification.

PAC Production Action Control.

PAPC Production and process control.

PCA Process capability analysis.

PCB Printed circuit boards.

PERT Program Evaluation and Review Technique. Similar technique to CPM. CPM and PERT differ primarily in their treatment of uncertainty in activity time estimates.

PFMEA   Process FMEA focused on the manufacturing steps, process flow, sequences, equipment, operators, test methods, preventive maintenance, and personnel training.

PHA   Preliminary hazard analysis.

PhRMA   Pharmaceutical Research and Manufacturers of America.

Plan   A method for achieving an end; detailed formulation of a program of action.

PLC   Programmable logic controller.

PM   Preventive maintenance.

PMA   Premarket Approval.

POF   Physics of failure.

Ppk   Process performance ratio.

PQ   Performance qualification.

PVC   Polyvinyl chloride.

Pyrogen   A fever-producing substance.

QA   Quality assurance.

QE   Quality engineer.

QFD   Quality function deployment, a technique for documenting overall design logic. It consists of a series of interlocking matrices that translate customer needs into product and process characteristics.

QS   Quality system.

QSIT   Quality system inspection technique.

R&D   Research and development.

RA   Risk analysis.

Reliability   According to ISO 8402-1986, the ability of a product to perform a required function under stated conditions for a period of time.

Rework   Action taken on a nonconforming product so that it will fulfill the specified DMR requirements before it is released for distribution.

RF   Radio frequency.

RPN   Risk priority number.

SCA   Sneak circuit analysis.

SFMEA   System FMEA focused on the customer and others involved (e.g., nurse, installer).

SHA   Software hazard analysis.

Six Sigma   Process improvement program based on the principles of DMAIC (define problem, measure opportunity, analyze process, improve, and control). This is also applicable for Design for Six Sigma.

SOP   Standard Operation Procedure.

SPC   Statistical process control.

SQA   Software quality assurance.

SWOT   Marketing tool used to evaluate the company's strengths (S) and weaknesses (W) and the opportunities (O) and threats (T) it faces in the market or industry environment.

TAAF   Test, analyze, and fix.

TCE Thermal coefficient of expansion.

Technology discovery and assessment Evaluation of new technology for applicability to a new product concept.

Testability The ability of a test method to test the results of a manufacturing process.

UL Underwriters Laboratories.

USP United States Pharmacopeia.

V&V Verification and validation.

WCA Worst-case analysis.

WCCA Worst-case circuit analysis.

REFERENCES

Abben, Richard P. 1999. "How Pacemakers Work and Why They're Needed." *Cardiovascular Institute of the South*. Houma, La.: Cardiovascular Institute of the South (URL: www.cardio.com).

Abernethy, Bob. 1996. *The New Weibull Handbook*. 2d ed. North Palm Beach: Abernethy.

AMC Safety Digest. 1971. *Fault Tree Analysis as an Aid to Improved Performance*. Washington, D.C.

ANSI (American National Standards Institute). 1992. *Software Reliability*. ANSI/AIAA-R-013-1992.

ANSI (American National Standards Institute). 1987. *Guidelines for the Verification and Validation of Scientific and Engineering Computer Programs for the Nuclear Industry*. ANSI/ANS-10.4-1987.

Asenek, V. A., M. N. Sweeting, and J. W. Ward. 1997. "Reliability Prediction and Improvement of Electronic Systems On-Board Modern Cost-Effective Microsatellites." *11th AIAA/USU Conference on Small Satellites*. Utah: AIAA/USU.

Bellcore. 1990. *Reliability Prediction Procedure for Electronic Equipment*. TR-332. Issue 6. Bellcore Customer Service, N.J.

CNET RDF 93. 1993. *Reliability Prediction Procedure for Electronic Equipment*. French Telecom.

Conformité Européen de Normalisation. 1997. *Medical Devices—Risk Analysis*. EN1441.

Deming, W. Edwards. 1986. *Out of the Crisis*. Cambridge, Mass: The MIT Press.

Department of Health and Human Services. 1991. "FDA Medical Device Regulation from Premarket Review to Recall." OEI 09-90-00040. Washington, D.C.: Office of Inspector General.

Donnelly, James H. Jr., and J. Paul Peter. 2000. *A Preface to Marketing Management*. 8th ed. Columbus: McGraw-Hill.

FDA. "Medical Device Use Safety: Incorporating Human Factors Engineering into Risk Management." Rockville, Md.

FDA. 1987. "Guideline on General Principles of Process Validation." May. Rockville, Md.

FDA. 1999. "FDA Enforcement Report Index—1999." (URL: www.fda.gov/po/enforcein-dex/99enforce.html)

FDA CDRH. "Human Factors Implications of the New GMP Rule." (URL: www.fda.gov/cdrh/humfac/hufacimp.html)

FDA CDRH. 1990a. "Device Recalls: A Study of Quality Problems." January. HHS Publication FDA 90-4235. Rockville, Md.

FDA CDRH. 1990b. "Preproduction Quality Assurance Planning: Recommendations for Medical Device Manufacturers." FDA 90-4236. Rockville, Md.

FDA CDRH. 1996a. "Medical Devices: Current Good Manufacturing Practice (cGMP) Final Rule: Quality System Regulation." *Federal Register*. 61 FR:52602-52662.

FDA CDRH. 1996b. "Do It by Design: An Introduction to Human Factors in Medical Devices." December 1. Rockville, Md. (URL: www.fda.gov/cdrh/humfac/doit.html)

FDA CDRH. 1996c. "Medical Device Quality Systems Manual: A Small Entity Compliance Guide." Chapter 3: design controls. December 1. Rockville, Md. (URL: www.fda.gov/cdrh/ dsma/gmp_man.html)

FDA CDRH. 1997a. "General Principles of Software Validation." Draft guidance version 1.1. Rockville, Md.

FDA CDRH. 1997b. "Design Control Guidance for Medical Device Manufacturers." March 11. Rockville, Md. (URL: www.fda.gov/cdrh/comp/designgd.html)

FDA CDRH. 1998. "Guidance for the Content of Premarket Submissions for Software Contained in Medical Devices." Rockville, Md.

FDA CDRH. 1999a. "FDA Reviewers and Compliance on Off-The-Shelf Software Use in Medical Devices." Guidance for Industry. Rockville, Md.

FDA CDRH. 1999b. "Medical Devices: Draft Guidance on Evidence Models for the Least Burdensome Means to Market; Availability." Rockville, Md.

FDA CDRH. 1999c. "Report on the Quality System Inspection Technique (QSIT) Study." April 26. Rockville, Md. (URL: www.fda.gov/cdrh/gmp/qsit-study.pdf)

FDA ORA. 1999. "Guide to Inspections of Quality Systems." August. Office of Regulatory Affairs Inspectional References. Rockville, Md. (URL: www.fda.gov/ora/inspect_ref/igs/qsit/QSITGUIDE.HTM)

GHTF. 1999a. "Design Control Guidance for Medical Device Manufacturers." June 29. Global Harmonization Task Force.

GHTF. 1999b. "Process Validation Guidance for Medical Device Manufacturers." June 29. Global Harmonization Task Force.

Hobbs, G. 2000. *Accelerated Reliability Engineering HALT and HASS*. New York: John Wiley & Sons.

IEC (International Electrotechnical Commission). 1998. *Analysis Techniques for System Reliability—Procedure for Failure Mode Effects Analysis*. IEC 812.

IEC (International Electrotechnical Commission). 2000. *Medical Electrical Equipment, Part 1-4: General Requirements for Safety—Collateral Standard: Programmable Electrical Medical Systems*. IEC 601-1-4.

IEEE (Institute for Electrical and Electronics Engineers). 1986. *Software Verification and Validation Plans*. IEEE STD 1012-1986.

IEEE (Institute for Electrical and Electronics Engineers). 1987a. *IEEE Guide to Software Configuration Management.* ANSI/IEEE STD 1042-1987.

IEEE (Institute for Electrical and Electronics Engineers). 1987b. *IEEE Standard for Software Unit Testing.* ANSI/IEEE STD 1008-1987.

IEEE (Institute for Electrical and Electronics Engineers). 1989. *IEEE Standard for Software Quality Assurance Plans.* IEEE STD 730-1989.

IEEE (Institute for Electrical and Electronics Engineers). 1990. *IEEE Standard for Software Configuration Management Plans.* IEEE STD 828-1990.

IEEE (Institute for Electrical and Electronics Engineers). 1991. *Guide to the Collection and Presentation of Electrical, Electronic, Sensing Component, and Mechanical Equipment Reliability Data for Nuclear-Power Generating Stations.* ANSI/IEEE STD 500.

IEEE (Institute for Electrical and Electronics Engineers). 1993. *IEEE Recommended Practice for Software Requirements Specifications.* IEEE STD 830-1993.

IEEE. Institute for Electrical and Electronics Engineers). 1994. *IEEE Standard for Software Safety Plans.* IEEE STD 1228-1994.

IIT Research Institute/Reliability Analysis Center. 1999. "Reliability Prediction." *START.* 4(2).

ISO (International Organization for Standardization). 1994a. *Quality Management and Quality Assurance Standards—Guidelines for Selection and Use.* ISO 9000-1.

ISO (International Organization for Standardization). 1994b. *Software Packages—Quality Requirements and Testing.* ISO/IEC 12119:1994.

ISO (International Organization for Standardization). 1997. *Quality Management and Quality Assurance Standards–Part 3: Guidelines for the Application of ISO 9001:1994 to the Development, Supply, Installation and Maintenance of Computer Software.* ISO 9000-3:1997.

ISO (International Organization for Standardization). 1998. *Medical Devices—Risk Management—Part 1: Application of Risk Analysis.* ISO 14971-1.

Juran, Joseph M. 1992. *Juran on Quality by Design: The New Steps for Planning Quality into Goods and Services.* New York: Free Press.

Kececioglu, Dimitri. 1991. *Reliability Engineering Handbook.* Vol. 2. N.J.: Prentice Hall.

Lusser, R. 1958. *Reliability Through Safety Margins.* Redstone Arsenal, Ala.: United States Army Ordnance Missile Command.

Medical Device Directive 93/42/EEC. Article 23, Annex I. (URL: www.europa.eu.int/comm/enterprise/medical_devices/guidelinesmed/baseguidelines.htm)

MIL-HDBK-217F. 1991. *Reliability Prediction of Electronic Equipment.* Philadelphia: Defense Printing Service. Out of print.

O'Connor, P. D. T. 2002. *Practical Reliability Engineering.* 4th ed. New York: John Wiley and Sons.

Porter, Michael E. 1979. "How Competitive Forces Shape Strategy." *Harvard Business Review* (April).

Rice, J. A., and V. Gopalaswamy. 1993. "Availability Estimation of a High Cost, Highly Dependable, Non-Markovian System." *IASTED International Conference on Reliability, Quality Control and Risk Assessment.*

The Silver Sheet. 1999, 3 October. Chevy Chase, Md.: F-D-C Reports.

Society of Automotive Engineers. 1994. *Potential Failure Mode and Effects Analysis in Design (Design FMEA) and Potential Failure Mode and Effects Analysis in Manufacturing and Assembly Processes (Process FMEA) Reference Manual.* Surface Vehicle Recommended Practice. J1739.

Underwriter Laboratories. 1994. *Standard for Software in Programmable Components.* UL1998.

Weibull, Waloddi. 1951. "A Statistical Distribution Function of Wide Applicability." *Journal of Applied Mechanics.*

Wood, Bill J. 1999. "Software Risk Management for Medical Devices." In *Medical Devices & Diagnostics Industry.* Los Angeles: Canon Communications LLC.

INDEX

Printed in the United States
by Baker & Taylor Publisher Services